INNOVATION TOOLS

The most successful techniques to innovate cheaply and effectively

By Evan Shellshear

7 Publishing

ISBN-13: 978-0-646-95646-6

Contents

Foreword

January 2012 was the month Eastman Kodak Co filed for bankruptcy, overcome by digital competition.

Its saving grace? Selling off of its photographic film and digital patents, consumer-printing products, photo kiosks, online picture-sharing and document scanning businesses, cutting its workforce in half to 8,500 employees and slashing US$4.1 billion in debt which ultimately left shareholders empty-handed. Only these drastic steps allowed Kodak to emerge from bankruptcy as a much smaller business with an entirely different focus.

Interestingly enough, less than three months later, Instagram, the photo-sharing app with just 13 employees and relatively insignificant infrastructure compared to the pre-bankruptcy Kodak, was acquired by Facebook for US$1 billion. This signaled a changing of the tide and an evening of the playing field.

Businesses that once required millions of dollars in start-up capital, could now launch from the comfort of one's home with a few hundred dollars. The advent of cloud computing, buoyed by fast, affordable internet connections changed the game, giving small teams with a relentless focus, a fail fast culture and a speed that large companies simply can't come close to replicating a distinct advantage in an exponentially changing world where cost is no longer such an inhibiting factor for small companies.

So how are big businesses responding to this shift?

At current rates of churn it is foreshadowed that more than half of the Australian S&P 500 index will be replaced in ten years' time and that one in five listed companies faces delisting in the next five years alone. The short answer, clearly is, not well.

Large organizations are built on cultural foundations and beliefs that cause decisions to embrace certainty, mitigate risk by analysis, avoid failure and take large bets, slowly. Executive leadership is finding it a very uncomfortable and difficult proposition to seamlessly transition to a landscape that requires them to do the exact *opposite*. The exploration of truly disruptive innovation requires that teams embrace ambiguity, mitigate risk by doing, embrace failure as a mechanism to learn and take lots of small bets, quickly.

"This is not what we learned in business school", we're told. No, it's not—but the world does not stand still.

With the evolution of corporate R&D, we've gone from ad hoc invention at the start of the 20th Century, R&D labs through to the late 90s, corporate venture capital in the noughties towards the establishment of incubators and innovation outposts today.

Why are companies establishing innovation outposts?

Because large organizations have been built to execute upon a repeatable business model, not to find a new business model. Performance and success metrics have long been geared towards favoring short term wins which fly in the face of the exploration and harnessing of disruptive innovation.

Because traditional metrics such as Net Present Value and Internal Rate of Return, R&D departments tend to focus on incremental improvements in order to protect and prolong their existing business models and revenue streams and optimize Return on Assets. While this might appease shareholders who play the short game, it comes at the expense of long-term sustainability and competitiveness. R&D can simply not keep up with the pace of exponential technologies and a leveled playing field. Just ask Blockbuster Video.

Creating innovation outposts helps companies adapt to the frenetic pace of innovation in order to successfully navigate the minefield that is disruptive innovation and help them move past simply stretching their existing S-curves towards catching the *next* S-curve.

But it is but one of *many* tools in the corporate innovator's arsenal.

In order to remain competitive, companies need to make disruptive innovation a part of their DNA, aligned with corporate strategy and integrated into business units—not merely an afterthought.

A tool is only as good as how you use it—and I commend Evan Shellshear on putting together this great book which provides the reader not only with an overview of the plethora of innovation tools that progressive companies are successfully using to grow new revenues but also relevant implementation techniques and case studies that are invaluable for today's senior executives and innovation managers. To stay ahead of the game, the tools in this book, when combined with complementary processes and mindsets, will make the difference between being the disruptor and being disrupted.

And we should know. At Collective Campus we have worked with some of the world's largest companies to help them implement the mindsets, methodologies and tools required to successfully explore new business models and disruptive innovation in this era of rapid change. Using the powerful tools and techniques presented here, we have taught some of the most bureaucratic and change averse teams how to become more flexible, fast moving and how to keep catching the next S-curve.

With that, it's fitting that I close with a quote from one of the world's most prominent leaders, a man who guided General Electric for 20 years in which time the value of the company rose 4,000%.

"If the rate of change on the outside exceeds the rate of change on the inside, the end is near." - Jack Welch

STEVE GLAVESKI

Founder and Chief Innovation Consultant, Collective Campus

Preface

It's the start of the twenty-first century and the innovation buzz is deafening. It commands the attention of everything—from mainstream media and popular culture to scientific journals. Innovation is now the driver of economies and the key to retaining a competitive edge. In its simplest form, innovation seems to have become an all-encompassing term for anything new which creates value or is simply cool and popular.

Given its importance, it's clear why people want to write about innovation. But what more could really be said?

Well, actually a lot. Of the many books focusing on the topic, from failing fast to managing innovation in firms, there always seem to be two parts missing:

- What are the best innovation tools to reduce risk?
- How can one best exploit such tools?

Many authors have written extensively on risk management methodologies, such as the *Lean Startup*. But what I and other people want to know is, what are the best tools for these frameworks? For example, you want to produce a Minimum Viable Product, but to get there you need concrete techniques to most efficiently develop it.

Another issue some people and companies have is that they simply have too many ideas which they would like to trial. If we are going to try and implement them all, then how can we do this cheaply and effectively so that we have not gone broke doing so and simultaneously have effectively tested each hypothesis?

To answer these questions, I looked at what techniques innovative people and companies use to lower risk and how they choose and implement them. While researching this, I simultaneously looked for the evidence backing up these choices. It has been a long journey beginning with my PhD studies eight years ago. This book is the result of that extensive research.

You won't just find the best strategies available for low risk innovation here. You will find curated content which should be interesting, cutting edge, and guaranteed to work. When filtering the available techniques, I focused on new ones which haven't been covered ad nauseam, whilst also making sure that they had been proven to achieve the professed goals.

Once I found these techniques, I took a step back and asked myself an important question: what was the point? The answer to this question was not what I expected.

The chapters are written so that they can be read in any order and are autonomous in their significance. There are connections between some chapters, but necessary terms appear in each so there is no difference when reading them in an alternate order.

The chapters chosen take different approaches to maintain reader interest. I varied the style of some chapters to be able to effectively deliver the message. For example, I approach the behavioral innovation topic with a style that is typical of the behavioral finance literature present in such texts as, *Why Smart People Make Big Money Mistakes.*

I have tried to write in a style which is engaging but fact driven. From personal preference, I like to see the story behind innovation, but I also want to know that what I am reading is based on fact and not opinion. Opinion is OK, but it does not constitute evidence.

When it comes to the facts, a level of detail was chosen which allows the desired story to be told without boring the reader with too much information. After reading a chapter, one should have a good idea about the techniques and when they are appropriate to use. If you want more details about how to carry out some specific part of the discussed strategy, then I have included more details in the Notes and References chapter. This serves as a useful reference point for someone seeking to take advantage of the listed techniques.

By writing this book I have also gained significant insights into what drives innovation around the world. I can now see the logical process of development of one innovation technique from another, e.g. the evolution from crowdsourcing to open innovation. These developments also fundamentally influenced my choice of strategies. I wanted topics in areas where I still see an open vista in front of the technique. That is, an opportunity for the reader to keep exploring and innovating. Perhaps even start the next big innovation trend.

I have been influenced by many of the great writers on innovation, of which there are too many to mention here. I have probably borrowed ideas and inferences from others throughout the book without consciously being aware of doing so. If that is the case and you notice an unattributed idea, then I hope you feel honored, and not annoyed, knowing that your message has successful influenced my work.

Enjoy!

Chapter 1

Why More Innovation?

Imagine you are sitting in a classroom learning a foreign language, just like thousands of other students across the country. The teacher utters a phrase in an unfamiliar vernacular which you struggle to understand. Then comes the ritual: someone is asked to translate what was just said. Like your fellow pupils around you, you crouch down and gaze around in an attempt to avoid the teacher's radar glare.

You quietly pray that some smart kid in the front row will answer the question first, but the eagle eye focuses on you. Without a chance to proclaim your sudden terminal illness, you are asked to translate. Choking on the words, you mumble something which sounds half-correct. To your surprise the teacher nods in approval and asks the usual question, "Now why was that so hard?"

So why didn't you jump at the opportunity to rise to the challenge and show your intelligence? Or your fellow pupils for that matter? We've all experienced this situation, so we all understand the difficulty. But what is the real reason for our apprehension?

Many explanations have been given as to why we play it safe in such situations. Everyone from psychologists to economists have opined on this. Research, however, has shown there are many factors in place. Most of them fear related. The fear of peer attention and looking too smart, fear of giving a wrong answer and looking dumb, fear of being the teacher's personal answering machine, etc.

Acquiring a new language is an extremely valuable skill in our globalized world. By answering the teacher's questions, you consolidate your knowledge of the foreign language and test the understanding you have gained so far. Such opportunities are the fast track way to better language proficiency. The benefits must outweigh the anxieties, so why hesitate?

Answering this simple, but all-important question is the common thread found throughout this book. The fears hampering our progress in learning a new language are comparable to the fears hampering our ability to achieve what we are capable of in other areas of life. As with acquiring a foreign language, something as unsophisticated as reducing fears can be a catalyst for achievement and innovation. But how do we reduce fears? Lower the risk.

By making results feel more certain and our dreams seem less risky, we are moving into a space where the impossible suddenly seems possible. And if there is one activity where risk plays the deciding role, it is innovation. Whether exploring the unknown, betting on a hunch or commercializing a risky technology; these speculative endeavors can all reside under the buzzword of innovation. So to drive more innovation, I propose reducing the risk involved as a means of lowering this seemingly fundamental execution hurdle.

The reasoning goes, if we could innovate without the fear of failing, of being seen as a loser by our peers, or losing money, then maybe we too would innovate more. Research has shown that given an equal chance of winning or losing a fixed amount, we consider the potential loss to outweigh the equivalent possible gains. Just as in foreign languages, if you didn't consider an activity or pursuit to be too risky, then you might launch yourself gung-ho into realizing your dreams.

But how can we reduce the perceived risks causing these fears? Well, what if you were given the choice of "having a go" without feeling dumb for failing, without feeling like a loser, or simply without any negative consequences? What if I told you there exists a multitude of powerful techniques to do just that and get you closer to your dream? A fearless way to innovate. This can now be your reality.

In this book, I will take you on a guided tour of ideas and technologies which promise to do just that.

They are ushering in new ways of being innovative with minimal risk. They are granting the people using them a novel means of testing ideas without sending them financially or mentally bankrupt. They are allowing people to pursue their dreams and test ideas before wholly committing to them. By doing so, people risk failing at a "hobby" rather than their career. How does that sound? A failed hobby is, after all, much easier to stomach than a failed career.

If innovation were like this for everybody, where you could simply muck around with the next Google, Tesla, or

Amazon, then unleashing this innovation potential in society could unchain the wildebeests of creative destruction, levelling the competitive landscape and ushering in a new era of human empowerment. With the techniques presented here, this is something we can all partake in.

Before we do, let's take a closer a look at what we're talking about and get everyone on the same page.

Invention, Entrepreneurship And Innovation

Most people consider the term innovation to mean the creation of something new and useful, but it depends on who you ask. I like to think of innovation as sitting in the middle of the see-saw ridden by invention and entrepreneurship. Innovation connects the novelty of the invention with the purpose and value of entrepreneurship. My inclusion of entrepreneurship may seem strange to many and as if it deviates slightly from the mainstream understanding of innovation but bear with me. By entrepreneurship I mean the activity of undertaking projects to provide products or services which society demands and hence finds useful and necessary. Often these activities have a commercial aspect involved. However, what is always common to entrepreneurship is a specific type of risk which is core to innovation. To differentiate these terms better, let's consider some examples.

Invention

The world is full of new or useful things. Of course, possessing one without the other is not what anyone would consider an innovation. For example, the "genius" who invented the splife (spoon with a knife edge) surely originated a new idea, however anyone with a drop of entrepreneurism coursing through their veins would ignore such a useless idea. Given that the quickest way of removing the splife is via your cheek, few people would consider it useful; calling it an innovation is as astute as investing in square wheels.

Entrepreneurship

Examples of pure entrepreneurship abound as well. Here, the goal is simply to provide something of value instead of introducing a new product or service. For example, Burger King could be seen as a purely entrepreneurial venture, at least initially. The pair who began that particular fast food chain came up with the idea after visiting one of the McDonald Brothers' original stores. And even when the chain was being restructured, they hired McDonald's executive, Donald N. Smith, to reboot their own foundation. Not that Burger King hasn't introduced any innovations, even in the early years, it's just that the initial goal appeared to be a purely entrepreneurial one.

Innovation

The fulcrum of the seesaw is innovation. True innovations include well-known items such as the typewriter, telegraph, telephone, personal computer and photocopier. Not only were each of these new (in some sense), but they were also immensely useful, which lead to their widespread adoption and success. When people think of the word innovation, they often think of era-defining, revolutionary products such as the telephone or Internet, but that is only a small part of the picture. Innovations take all forms from disruptive technologies to equally important, small improvements of existing products or services.

For many innovations, the revolutionary step was not to invent something completely new, but merely to add a final touch or new feature to an extant idea. Such small, incremental innovations can lead to massive successes, e.g. the personal computer. Our desktop friends became the natural consequence of shrinking the room-filling mainframe computers of yesteryear. Even a major break-through like the Internet was only the end result of a long technological march towards its final widespread and dis-ruptive acceptance. The power of the Internet was un-leashed after many individual parts fused together, from software, to hardware, to new business models.

In fact, being the first to launch a genuinely new or dis-ruptive idea can be a disadvantage. Despite Xerox being the first company to invent the personal computer as we know it, they were the last to profit from it. Xerox endured IBM, Apple and a host of other companies run off with the winnings. It may seem unexpected, but often adding just an incremental innovation to an already established idea can catapult one to success, as is shown by products such as smartphones.

This is not to say that incremental innovations are more important than disruptive ones. One just needs to see the whole spectrum of pioneering activity to recognize and foster it in their day-to-day life and organizations. Small improvements can be just as successful as large ones (depending on how you measure it!) and we will constantly be presented with opportunities for both. Hence, our toolbox should contain strategies to deal with incremental as well as disruptive innovation.

Everyone Is Unique

So what is the elixir of courage that can drown the natural risk aversion in us? Unfortunately, there is no one magic recipe, because everyone is unique. For example, it has been shown that people's innovation threshold varies wildly. This is not just about people's ability to innovate, but especially their willingness to innovate. Robert Sternberg, former dean at Tufts University and a leading creativity scholar, claims that, "creativity is in large part a decision that anyone can make but that few people actually do make because they find the costs to be too high."

So, according to Sternberg, if we simply create the right conditions, we can spur innovation along. Innovation experts Paddy Miller and Thomas Wedell-Wedellsborg write in their acclaimed *Innovation as Usual* book, "what is key about these people is that they won't ... [innovate] if the personal cost of persistence is too high."

And what personal cost is higher than failure? One of the key conditions to promote the desired behavior is clearly a low risk environment powered by low risk innovation tools.

Although different people have different risk affinities, you may not know that your own appetite for risk changes, based primarily on what you are betting on. For example, research has shown that we tend to be more risk averse when gambling for a gain than when trying to avoid the same loss. Hence, even though you may think you always behave in the same way regardless of the situation, this isn't the case.

But it is not just for losses and gains that we change our risk profile. A racing car driver pushes his car to the limit on the race track, risking his life and that of those around him. When he drives home in his own car though, he may not even be willing to drive through an orange light, i.e. he is completely risk averse. Scientist have even been able to demonstrate this type of behavior in rats by altering the expression of certain genes in their brains which means the changes are inherent in the brain and so capable of occurring given the right circumstances.

So you may look at your favorite entrepreneur and think all that it requires to get your idea off the ground is to be less risk adverse. This belief is wrong. Each situation is unique and presents a level of risk that may even scare off the most battle-hardened hazard hustlers. So searching for cheaper, less risky ways to innovate is the perfect strategy to demolish the psychological barriers of entry into the world of innovation.

And the remaining few who are risk seekers in everything they do? Well, wouldn't it be worthwhile to minimize the damage each loss does and so become more successful? Especially if doing so was cheap and easy.

Innovating On The Cheap

Being innovative is to wager your mental horsepower against the world. It is you and your idea against everyone else. Such a challenge is daunting, to say the least, so help is welcomed by most sane innovators.

We know that innovation can be expensive and has a high chance of failure. Everyone has heard the maxim that 90% of all entrepreneurs fail. So where does that leave these risk takers?

Well, depending on how you measure success, the reality is worse. It has been shown that as few as 7% of all innovators successfully commercialize their inventions. Other studies have shown that 4% of innovation initiatives achieve their internally defined success criteria. Only 12% of research and development projects even return their capital cost.

So why would anyone bother innovating? Given that the odds are stacked against you, why would you devote yourself to something which is so likely to fail? Sure, we all play things like the Lotto or enjoy the odd bit of gambling, but that's only a few dollars (in most cases!). So that an innovative concept becomes reality, it normally involves a significant expenditure of our time, effort and money.

For companies, the purpose of innovation is clear. A focus on innovation and making it a part of corporate culture leads to increased profits, market value, survival in the market jungle and much more. But because companies are operated, of course, by people (at some level), it is people who must take the risks and make the decisions. Hence, in certain situations, commercial expediency forces many corporate executives to take on the risks of innovation and learn to deal with them. As the methods in this book show, what types of risks and exactly how they are taken can be significantly influenced by the individual.

When the right techniques are used, they can mean the difference between a cheap experiment and an expensive failure.

If we zoom out and examine many companies' and startups' innovation strategies, we notice two major low risk strategies often employed by organizations worldwide when launching risky and innovative products or services:

1. Copy a successful, similar product or service and adjust it to the current problem.
2. Incrementally innovate towards the final goal via a "lean" or "agile" methodology, checking and testing the idea along the way to minimize risk and abandon detrimental ideas as early as possible.

These strategies minimize risk by working with frameworks which have proven to be successful and, in the first case, the strategy is a well-known and understood one. The second paradigm has become very popular in the last 20 years. Instead of going "all in" on one or two big ideas, the goal is to be able to place many bets by failing cheaply and fast so that you can keep churning out and testing new ideas.

This method of innovating is one of the driving forces behind the new innovation tools presented here. Ways of cheaply testing ideas and therefore lowering risk will always improve your innovation chances by weeding out the bad ideas and allowing your good ones to blossom. And in fact, this leads us to the next point.

So What About The Other Stuff?

Fundamentally, risk underlies all innovative processes and successful innovation means managing this risk appropriately. A good first step is to adopt a risk management perspective to innovate in a low-risk way. These frameworks usually contain best practice processes, but what is almost invariably missing is a set of specific methods to achieve the desired process outcomes. The tools presented in this book will provide you with the meat on your framework bones to implement your low-risk management strategy.

For example, the ISO 31000 risk standard provides guidance on the management of risk in almost any situation. Its publication by the International Organization for Standards in 2009 was a milestone. The structure and principles build on the extremely successful AS/NZS 4360 risk management system developed in Australia and New Zealand. According to the standard, risk is seen as the "effect of uncertainty on objectives" and the management of it is defined as "coordinated activities to direct and control an organization with regard to risk."

A critical part of defining risk is the word objectives. Objectives vary across industries and even within an industry. They can include minimizing cost, minimizing damage to reputation, minimizing risk of ineffective processes, but they could also be very specific, like building a smartphone.

At the core of the global standard is a risk management framework, which outlines powerful principles and also demands risk mitigation strategies to reduce the risk associated with achieving a given objective. When assessing the risk of your objectives, you will need to find the best possible tools to mitigate the risks you identify. Such tools are the muscles that move your entire risk framework and also the focus of this book.

Those who are involved in the startup scene should be aware of an extremely popular framework for risk management known as the Lean Startup. Popularized by Eric Ries in his book *The Lean Startup*, its management principles focus on powerful objectives for new product and service development, in particular for new companies attempting to build a business around their latest offerings.

The centerpiece of Ries' strategy is to focus on learning and testing hypotheses, both in products and organizational efficiency, as quickly and cheaply as possible to minimize waste. Startups should focus on those things which lead to validated successes and bring a startup closer to its goals. At the same time, this should be done in a way that incorporates customer feedback as early in the development cycle as possible, and so focuses on concepts like a Minimum Viable Product for customers to test. It is a powerful process used by both tiny startups and, perhaps surprisingly, large multinationals worldwide.

The Lean Startup methodology (like the ISO Standard) is process-focused, rather than content-focused, so you need to figure out for yourself what content you require to achieve your objectives. By having a clear focus on your goals and where you want to go, my book will provide you with the vehicle to take you there. By the end of this book, you will be able to fill some of the world's most successful processes with the world's most successful content.

My Innovation Agenda

I will take you on a journey to master six of the major tools to minimize your innovation risk that are available right now. These six methods are really toolboxes containing many techniques, all proven to significantly improve the delivery of your breakthrough ideas by maximizing your chances of success.

We will closely examine everything, from how you can use the minds of the world for your own purposes, to ways of cheaply and effectively exploiting capital-intensive tools. The choice of topics was driven by methods known to work and demonstrated to reduce innovation risk. There's no unproven theories or esoteric philosophy here.

The strategies presented in this book are not the only ones available to reduce your innovation risk; they are simply some of the most powerful. Other successful methods to lower risk, such as technology brokering, have been demonstrated by none other than one of the world's most impressive innovators, Thomas Edison.

The idea behind technology brokering is to lower the cost of innovation by combining known technology—"off the shelf" components—to create a new solution. The so-called Wizard of Menlo Park used to say that to innovate you need a good imagination and a pile of junk (not a pile of expensive things!). In other words, innovation can and should be done cheaply.

Edison not only talked the talk, he also walked the walk. When he was developing his first functioning light bulb, he initially designed an experimental version in which he could easily replace the filament to test many possibilities. During this phase Edison tested horsehair, cork, rubber, grass fibers, wrapping paper, fishing line, silk and even the beard hair from his laboratory workers. His choice of material was guided by a systematic attempt to test easily acquirable and cheap products. This meant that as soon as one of his trial materials was found suitable as a filament, not only was the cost of his experiments kept down, but ipso facto the later production cost too. This is an important lesson we can all learn from.

Another area of new low risk tools and techniques not addressed in this book is talent acquisition. It is here that we are also seeing significant risk-shrinking developments which are making it simpler to access the best people through a host of software and online tools. Websites such as LinkedIn, UpWork and YourEncore are some of the biggest service providers in this regard. Being able to easily find and scale up human resources on an as-needed basis, in a similar way to software (examined in a later chapter), is a massive boon. It reduces the risk of committing to employing thousands of people for unproven products. Not to mention the time and money saved from not having to filter candidates or hire expensive recruiting agencies means startups (and even large corporations!) can focus their limited resources on their core objectives.

Technology brokering and talent acquisition are some of a number of methods to reduce innovation risk which are not considered by this book. While important, these methodologies were left out because they are a less than perfect fit with my choice of topics.

Of course, there's only so much room in a single book. Naturally, there are a few sub-topics I do touch on which could be explored in more depth. Sometimes that's because the technique is new and its usage is not as clear as I would like, or the application is rather straightforward and self-evident. Whenever possible, I've included references to provide you more "nuts and bolts" details.

For example, one topic which could be discussed further, is the wave of new online learning platforms such as Udacity, edX and many more. Companies such as these are lowering the barriers of knowledge acquisition. This in turn complements several techniques presented in the following chapters. By being able to cheaply and easily access the requisite knowledge for a job, many unnecessary pitfalls can be avoided. However, at the time of writing, the industry was still in a state of constant flux with many seemingly good business models failing and new, more promising ones appearing.

Another topic, only addressed in a tangential fashion in some of the chapters, is simply the access to customers. Whether via new marketing techniques such as AdWords, partnerships with e-commerce companies (Etsy, Amazon, etc.) or physical stores (Quirky, Target, Best Buy, etc.), there are many opportunities lowering the innovation risk for companies. Especially by making the launch into new and competitive markets simpler and more effective. These services and tools are often able to provide customer intelligence and detailed analytics, thus making a successful product launch that much more likely. AdWords and other such tools are covered quite extensively in a number of other resources, so it seemed unnecessary to repeat their best practices in this book.

So What Will I Get From This Book?

This book will reveal to you the best tools available to drive your innovation agenda in the lowest risk and most effective way possible. Each technique will be defined and then examined in detail via numerous case studies to allow the reader to find analogous situations to the one he or she is currently facing. Finally each chapter will draw a conclusion outlining some final considerations which should be borne in mind when using the methods expounded therein.

Beyond these practical benefits, by the end of the final chapter, you will also understand why low risk innovation is more important than simply making innovations more likely to succeed. At the conclusion of the book, I will introduce you to the virtuous cycle of innovation and provide some evidence as to why I believe my hypothesis about virtuous innovation cycles is true. So no matter what your goals are, the overall benefits you derive will probably be bigger than your initial goals and lead you down an unexpectedly rewarding path.

Before We Begin

A final important note to remember is that even with an innovation toolbox of effective, inexpensive tools, there is still a lot of luck involved when innovating. Luck—that is, everything else we can't quantify or currently explain—plays a big role in uncertainty and risk. Part of this can be mitigated with the tools presented here, but part of it is simply out of your control.

When two young budding entrepreneurs tried to sell their business to Excite for $750,000, the CEO of Excite rejected the offer thinking the company was not worth the money. With few other options, the young businessmen kept slugging away. Today, they control one of the giants of the internet world, Google. Had they sold their business for $750,000, who knows whether Google would be the household name it has since become.

I hope to inspire you to search for cheaper and less risky ways to innovate. Especially, to help you in your endeavors so you don't end up exhausting your resources on one idea. By looking at the techniques others have found to lower their risk, we can hopefully draw parallels with what we are doing and raise our chances of success. As more people start using the techniques presented here, the number of innovators who fail will be less likely to be 90% and more likely to be 9%.

Remember, the secret to innovation is not to avoid failure, but to keep your failures cheap and to learn from them. For most of us then, with less time and money invested in our innovations, our successes and failures will seem like they are part of an enjoyable hobby that we do in our spare time. And who ever really cared that much about failing at a hobby? You can always find a new one.

Chapter 2

Share The Risk And Increase The Gain

The humble bike is a great symbol of the masses. It's an empowering tool that provides people around the globe with a faster and more efficient form of mobility than a pair of feet. But this mode of transport has done much more than just accelerate our passage from A to B; it has also turned the wheels of major historic events. From early women's liberation to massive sporting challenges, the bicycle provided a crucial mechanism to reach humankind's goals.

It was upon this two-wheeled device that a young entrepreneur was compelled to search for a way to solve an annoying problem. And in finding a solution, he would invent a product that would open a landmark chapter in the recent history of crowdfunding. It was an event that helped push crowdfunding into the attention of your average citizen and cemented its credibility as a new way of financing risky ventures. But his story must wait a little while longer because this new form of backing from the masses is only the beginnings of something much bigger which we should address first.

Crowdfunding is part of a wider movement in society whereby large numbers of regular individuals can support projects and ventures by pledging relatively small monetary contributions, rather than relying on a few large traditional investors—a powerful risk reducing innovation technique. If we take a step back and see the bigger picture in which crowdfunding is operating, then an even more disruptive phenomenon comes into view: crowdsourcing.

Crowdsourcing

The word crowdsourcing was coined in 2005 by two editors at Wired Magazine, Jeff Howe and Mark Robinson. It arose after the pair noticed a proliferation of businesses using the Internet to outsource work to the public at large. According to Jeff and Mark, a crowdsourced activity is one where members of the public are sought to contribute to a project by giving their time or money. Sometimes their contribution is compensated, and sometimes it is done simply out of good will.

Even before a name had been found to baptize this phenomenon, crowdsourcing was already a major trend. During the early 2000s, many companies embraced the crowd spirit to launch new ways of doing business, charity work and almost anything which could be divided up into smaller parts doable by many individuals.

Wait! This must be the ultimate capitalistic business model, right? Devise something useful which an individual cannot finish by his or herself. Create a platform to realize this goal and then crowdsource the work to millions of faceless volunteers to complete the job. Free, or at least cheap employees and limitless profit!

This could not be further from the truth. Successful crowdsourcing is difficult and fraught with problems. To engage a crowd usually requires one to appeal to something else in people other than their desire to earn money. Even when offering compensation, surprisingly, people will often rather volunteer their time than accept a small payment to help others.

In an example described in the landmark book, *Predictably Irrational,* written by the renowned behavioral economist Dan Ariely, a group of lawyers were offered $30 an hour, instead of their usual far higher fees, to work for impoverished pensioners. The fairly obvious result? They refused. However, in an unusual twist, the lawyers were then requested to work on a pro bono basis, and they promptly agreed. Hence, the way to garner people's willingness to perform work is not as obvious as our intuitions may lead us to believe.

However, the volunteer vs. paid conundrum is typical of many decisions facing beginners wanting to crowdsource a solution to a problem. So, if we decide not to engage people on a voluntary basis, and there is a task which one can put a price tag on, then why shouldn't one outsource the job to the "crowd" and wait for the results? The first problem that arises is the deluge of ideas. The many, many suggestions mean someone has to consider and evaluate them.

It may or may not be a good way to solve a problem— to dedicate resources to assess the crowd's responses to a task. Those same resources may be better used solving the problem itself, instead of sifting through the half-considered, half-baked idea-dough from amateurs. Especially if the person reviewing the submitted ideas is a highly trained specialist.

However, in reality many companies assign the task of sifting through responses to junior staff to save money. Unpaid interns are often used for this task, although it leaves one wondering how many good ideas are shipwrecked in the shallow experience of the corporate ordinary seamen?

One way of dealing with this problem is to reemploy the crowd. Ask the crowd for the ideas and then ask them to vote for the best one! For certain activities this can be a useful trick. However, just because people are able to suggest solutions does not guarantee they are capable of successfully evaluating them. We will return to these questions later on in the book and see a number of examples of how to successfully address these issues.

Despite these difficulties, to say that crowdsourcing has been anything less than a revolution would be an understatement. The many challenges which have been conquered by this method are impressive. One which most readers will be familiar with and have probably used is Wikipedia.

The effect Wikipedia has had on the lives of net citizens is huge. The "wiki" is one of the top ten most visited websites in the world, has nearly 35 million articles in 288 languages, and 500 million unique visitors each month.

Almost twice the population of the United States of America visits the website every month! Like most Internet startups, it has achieved all this at the young age of only 15.

Wikipedia is one of the largest crowdsourced projects the world has yet seen. There are more than 19 million accounts registered, almost the size of Australia's population, and as of November 2014, there were around 70,000 active editors.

These contributors are not only producing one of the world's biggest repositories of information and knowledge, but at the same time facilitating another revolution touched on in the later chapter about free knowledge. Not only are contributors showing the power of crowdsourced work, but they are also providing fodder for the open access revolution. Wikipedia entries are one and a half times more likely to refer to open access articles than closed access articles, thereby they are also raising awareness of, and support for, the laudable open access efforts. We will return to this theme later.

The citizen's encyclopedia is a prime example of a crowdsourced, cost-cutting innovation. Its rise to prominence has helped throw crowdsourcing into the limelight. Despite the novelty bestowed upon crowdsourcing by the Internet and projects like the giant Internet encyclopedia, its basic principles have been around for centuries. An early example of crowdsourcing is the definitive Oxford English Dictionary.

To create this cornerstone of the English language, back in the 19th century volunteers from all over England scoured all available books and recorded the earliest usage and meaning of words in all their possible contexts. This carefully collected information was sent to the masterpiece's editors, who then compiled the words into the final set of books. Without the Internet to help the flow of information, the completed work took a little longer than its online cousin did to reach a similar size—more than 70 years!

The reason for our current interest in crowdsourcing is in what it means for innovation. Crowdsourcing allows one to spread the risk of an enormous task among thousands of people with the possibility of avoiding the financial liability of such workers. The added benefit is that if disruptive innovations require disparate perspectives, crowdsourcing is a great way to achieve that.

In addition to obtaining crowd help with various activities or projects, easy access to the public at large can also be used to gauge its reaction to a new idea or product. This readily available feedback can help prevent one from plowing millions of dollars into something that is likely to be a flop. This idea in turn has led to the rapid rise of crowdfunding. Its dramatic effect on social ventures and innovation is only starting to become clear.

Crowdfunding

As a portmanteau word, "crowdsourcing," has become so diverse that many subcategories have arisen. For example, crowdcasting, crowdfixing, crowdtesting, as well as another major development important for this chapter, known as crowdfunding. Crowdfunding turns the crowdsourcing of work on its head. Instead of the crowd doing the work, you do the work, and the crowd pays.

Crowdfunding is a way for people to back ideas they believe in, simply think are fun, or just because they want to. Platforms exploiting the mass funding model have come to the fore in our Internet connected age because of the ease with which one can connect with millions of financial sponsors.

If you can convince one person to part with one dollar, then that is all you have, one dollar. If, via leveraging the Internet, you can convince one million people to part with one dollar, then who needs to court banks and rich investors? This is the principle behind crowdfunding. It does not, of course, stop people from contributing more. In many crowdfunded projects, there are people who may contribute tens of thousands of dollars merely because they admire an idea and want to see it realized. If you do part with thousands of dollars to help finance a project, then apart from receiving your own signed version of what you supported, it is typical to receive perks such as meeting the inventor in a personalized gift ceremony.

Currently, there are thousands of projects on numerous platforms soliciting crowdfunding. A quick Internet search revealed more than 35 such platforms operating at the time of writing. This provides a plethora of choices and may actually make your crowdfunding project more difficult to start than you thought! Despite the large number of platforms, there are two major ones, Kickstarter and Indiegogo, which seem to dominate the rest.

Kickstarter

Located in Brooklyn, New York, Kickstarter began operation on 28 April 2009. It was founded by three artistically oriented entrepreneurs, Charles Adler, Yancey Strickler and Perry Chen. Kickstarter was Perry Chen's baby, having mulled it over in his head for years while he was struggling to get his music career off the ground. He met Strickler in Brooklyn and, being the only person he knew who had Internet skills, he suggested his idea to him.

Charles Adler arrived later, in 2007, as a co-founder and joined because, according to him, "[T]he vision was so compelling." But it didn't take long for others to realize this too. Already in 2010, Kickstarter was named one of the "Best Inventions of 2010" and the following year, one of the "Best Websites of 2011." The platform has grown strongly over the years, expanding to the UK, Canada, Australia and New Zealand in 2012 and 2013, and then to Northern Europe in 2014. To date, Kickstarter has achieved the highest funded projects for any crowdfunding platform.

To launch a campaign with Kickstarter, you need to fit into one of the 13 categories allowed, be a creator, and abide by the prohibited uses such as charities and awareness campaigns. If you are going on to produce hardware like a new smartphone, then you have additional requirements such as evidence of a physical prototype and having a manufacturing plan. All of this is designed to give people the feeling they are backing serious, well thought-out projects and not just ordering products from Amazon.

Kickstarter backs projects in a wide range of categories. However, of all the categories, Film & Video, Music and Games account for over half the money raised. In fact, Games raised two out of every ten dollars pledged on Kickstarter in 2013, whereas about 10% of the films accepted by the 2012 Sundance, Tribeca, and South by Southwest Film Festivals were funded on Kickstarter. The quality and number of successful ventures going through this funding platform are significant.

Indiegogo

The other major crowdfunding site, with headquarters located on the other side of the U.S. in San Francisco and only operating in America, is Indiegogo. It was one of the first websites to offer crowdfunding and was started by three people experiencing a similar difficulty. All three founders shared a passion for the idea of democratizing fundraising, having had experience in industries where it was difficult to fund great projects that may not return an obvious gain, usually financial, to backers.

Its creators, Eric Schell, Slava Rubin and Danae Ringelmann launched the website in 2008. Since then it has also grown considerably, even partnering with President Barack Obama's Startup America to offer crowdfunding to American entrepreneurs.

Although both websites use a crowdfunding model, they differ in the details. If a Kickstarter campaign does not reach its funding goal, then all funds are returned to the backers. If an Indiegogo campaign fails to reach its announced financial objective, then all funds collected (minus Indiegogo's fees) are still given to the creator of the project. Although some projects on the West Coast startup's platform can still be chosen as an "all or nothing" fundraising and will return funds to backers if a project fails to reach its stated goal.

Apart from being more relaxed about the distribution of proceeds, Indiegogo is also less restrictive than Kickstarter as to who can request funding. You can reach out to fund a cause, charity or even a startup business. So for those looking to avoid Kickstarter's stricter rules and project categories, Indiegogo is probably the preferred option.

Between them, Kickstarter and Indiegogo have launched somewhere in the order of 300,000 projects at the time of writing, with just under 100,000 of those on Kickstarter reaching their goals and very likely with a similar number on Indiegogo. Altogether, the fundraising activities have collected somewhere in the order of $3 billion dollars to back ideas from around the entire world.

These projects cover just about everything, ranging from comic books, animated features such as films and TV shows, to games and even hardware. One such notable hardware project was the Pebble Smartwatch, which helped to usher in widespread acceptance of a new category of connected devices, the smartwatch.

The Pebble

It was the year 2008, and people worldwide were still excited by a revolutionary device which had exploded onto the mobile market the previous year. The device was the smartphone, and not many years later, the majority of all Americans would own and use one. In fact, one billion smartphones were already in use worldwide by 2012. The smartphone had changed people's lives, but it also had a big problem.

The new Internet phone connected and enriched our lives like no other mobile consumer device. We could now receive emails, SMS's, telephone calls, and all sorts of notifications from wherever we liked. However, this very same information assault suddenly made things like riding a bike or driving a car very dangerous.

Many countries are oblivious or slow to respond to the issue of whether driving and talking on a mobile is safe or not, but the facts are unequivocal. Interacting, talking, SMSing, etc., while driving is a leading cause of serious accidents around the entire world. You are four times more likely to crash your car while using your mobile than not. This is about the same as downing two standard glasses of wine then jumping behind the wheel. It's not hard to imagine that similar dangers exist for bike riders.

This was the situation confronting a young University exchange student, Eric Migicovsky, studying abroad at the Delft University of Technology in the Netherlands. As with many other young tech enthusiasts, he owned a smartphone. And like most of the Dutch, he had a bike.

Riding a bike in the Netherlands is one of most convenient ways to navigate its great plains. The country is extremely flat with almost a quarter of all the land area being at or below sea level. In fact, to make fun of the lack of hills while riding a bike, some Dutch call a headwind a "Dutch Hill."

However, as he was of Canadian origin and not from the Netherlands, the soon-to-be Pebble Inventor had not had the same opportunity to master the art of riding and texting. Eric dreamt of being able to see what was happening on his smartphone without having to take it out of his pocket while riding. Responding to constant notifications and messages was a dangerous habit for his untrained Canadian hands, which, at the very least, could easily fumble a phone and drop it while riding.

So Eric came up with a solution: move those notifications one meter out of his pocket and onto his wrist.

In 2008, he put together the first prototype in his university dorm. His creation, to express his idea of the biker-friendly smartwatch, was a little Arduino circuit board and a Nokia phone screen. He even posted a video of it on YouTube to let the world see his invention.

Motivated by the idea and the encouragement he received, he continued working on his watch with some friends at the University of Waterloo. After receiving positive feedback, including indications of interest in buying such a device, he formed a company, Allerta, and went ahead with production.

This wasn't the first time someone had thought of making a smartwatch. In fact, at the time, there existed several similar products. One of the first major consumer-oriented plunges to put a computer on our wrists was by Microsoft back in 2003. Even in 2008, you could buy the

Fossil Wrist Net smartwatch, but at almost $400, it was out of reach of most students.

Eric's next prototype was a product called InPulse. The product was a first iteration and featured only the basic watch functionality. It wasn't waterproof, had a seven hour battery life, no touchscreen and could only run one app at a time. It worked with Blackberry and Android, but it could not be connected to the iPhone because Apple did not support Bluetooth until after it released iOS 4. It was a milestone for the young entrepreneur.

However, InPulse was also a complete failure and almost cost Eric his company.

It was 2011, and to produce the first batch of soon-to-fail devices the young Canadian entrepreneur was able to raise around $350,000 from angel investors. It was a risky venture that exhausted almost all the company's cash resources to manufacture a large quantity of smartwatches. As stated, they flopped. Only 50% of those who pre-ordered watches followed through with their order, leaving Eric beneath a mountain of unwanted devices and with almost no cash.

In spite of this failure, the team learned a lot. They discovered that notifications were the key, as well as a readable screen in all light conditions, and a long battery life. Having a waterproof device was also desirable because people wanted to use it like a normal watch. Finally, cross platform support and an always-on display would be the last two features, which they hoped would ensure a winner.

So they changed the name of the company from Allerta to Pebble and got to work on these key selling points. However, just having the right specifications would never be enough. Eric realized that for Pebble to truly succeed, he needed to get some excitement behind the product launch, and this is why he turned to Kickstarter.

The Kickstarter campaign was launched on 11 April 2012. To initiate the process, the Pebble team had set a modest goal of raising $100,000 to fund the development of the first prototypes. What happened next was beyond anything anyone could imagine.

Within only two hours of going live, Pebble had smashed the $100,000 goal. Within six days, they raised over $4.7 million and became the biggest funding success in the history of Kickstarter. And there were still 30 days to go!

In the end, between 11 April 2012 and 18 May 2012, the campaign raked in $10.3 million. No other product had ever raised this much on Kickstarter before. And the successes only continued. When Best Buy began to sell the Pebble smartwatch in July 2013, all models were sold out within five days. By the end of 2014, the one millionth watch had been sold.

Not content with their success, on 24 February 2015, Pebble announced a new watch to add to the growing smartwatch range, the Pebble Time. They decided to utilize Kickstarter again to support their new endeavor. If the Pebble campaign was successful, then the Pebble Time campaign was about to redefine the word success.

Consistent with their previous campaign, within 17 minutes they raised $500,000. A mere 32 minutes later, they breached the million dollar mark, as well as creating another record for the crowdfunding platform. The money flood continued demolishing records by raising $10.3 million in 48 hours. By 3 March, it was the highest funded Kickstarter campaign ever, with $14 million dollars already pledged. And again there were still 24 days left.

Eventually, the Pebble Time campaign raised $20 million. Needless to say, this was also a record and further proof of the crowdfunding model's power. Critics of the Pebble Time campaign claim that the Kickstarter model should only be used by budding new enterprises, but Pebble showed that this method of spreading the cost of innovation can even be used for a more mature company's further product development aspirations. In any case, the first Pebble was also not Eric's first smartwatch, it was just his first well-known one.

The Crowdfunding Revolution

The crowdfunding model has changed traditional funding paradigms. Where more traditional financiers might be reluctant to finance a product that itself may not be successful, or that there may not be any demand for, crowdfunding provides a simultaneous solution to both issues. A single source of finance no longer has to carry the entire potential loss for an ill-conceived product that no one wants. Crowdfunding is truly one of the most powerful techniques examined in this book for reducing innovation risk.

Platforms similar to Indiegogo and Kickstarter have also sprung up to fill other funding gaps not covered by the two major players or to specialize in a specific aspect. Websites such as GoFundMe use the crowdfunding model to realize inspirational goals. From helping finance top students unable to pay for their college education, or paying unaffordable medical bills, according to GoFundMe, websites such as these have the goal to fund "inspiring campaigns by incredible people."

However, crowdfunding can have a negative side for the layman investor. Because the funds are to be used to create products which do not yet exist, there is a large inherent risk in funding Kickstarter products, in that the backer may not see a result. There can also be extensive delays in delivering goods, and some campaigns seem to be genuinely questionable. For example, in 2011, Matias Shimada raised almost $2,000 for a new film but instead of making a new film, he simply plagiarized another one. However these problems are not unique to crowdfunding. What is different, is that typically the financial supporter can do little about it.

In certain cases, the behavior can be criminal, with the American Federal Trade Commission even launching investigations into a crowdfunding charlatan, Erik Chevalier, who refused to repay the promised funds after his board game project failed. He was eventually ordered to repay the money by authorities but claimed he was unable to.

In addition, the question in the back of every investor's mind is, "What will my money be used for?" If it is to develop the product they are backing, good. If it is partly for the creator's holiday in the Caribbean, bad. There have been cases where creators appear to have been overly generous in the expenses for their campaign, which has raised questions regarding the use of the backers' money.

Then there are those projects which seem legitimate, but straddle the border with scams. The Anonabox is one such project which caused a revolt in the ranks of the Kickstarter community.

The Anonabox was a product which promised to give users online anonymity and thereby prevent censorship. The Kickstarter protocol states that campaigns need to be for "products in the making" and not simply a consumer product. In the case of the Anonabox, backers started crying foul when they discovered that the Anonabox team had bought a readily available product and only slightly tweaked it. Others also questioned the security it promised, with Wired writer Andy Greenberg alleging, "it could create more risk than protection." At the height of the vitriol it generated, more money was being pulled out of the campaign than pledged. After less than a week, the project was suspended.

However, the story did not stop there. The project's creator then proceeded to launch the project on Indiegogo, aiming for the modest amount of $20,000. And despite the bad press surrounding the Kickstarter campaign, he easily surpassed that goal and raised $82,742; still, a far cry less than the roughly $600,000 pledged before the previous campaign was challenged and cancelled.

This project is one of many which have been cancelled on Kickstarter due to unfounded price or performance claims, plagiarizing parts of the campaign from elsewhere, or for equally unsavory reasons, and it seems they made the right decision.

Even though Anonabox was successfully relaunched, allegedly having learnt from its previous experience, the initial batch of devices was revealed to have a major security flaw and was promptly recalled. This flaw surprisingly concerned a basic issue which any company touting privacy as its selling point learns on day one; password security.

Growing pains aside, whether it is a project to revolutionize a product category or funding to save someone's life, the crowdfunding model has turned financing on its head. The control of investment money was once mostly dominated by two institutions: venture capital (in its many forms) and banks. This has now been overturned by people merely looking to back ideas and projects they think are interesting and worthy of support.

It also gives entrepreneurs a new avenue in which to raise funds and adds competition to the money market, which may help innovators acquire finance on better terms. Best of all, it provides funding for those with whacky ideas, such as a crowdfunded potato salad!

However before you enter "crowdfunding" into Google in order to find your next source of investment, a few words of warning. Putting together a successful campaign to achieve your financial goals is not easy. It can take months of work and quite a bit of luck to capture the requisite media attention for your idea. Often high quality videos, detailed product breakdowns, testimonials and much more are essential for a successful launch. Books have even been written to help people crowdfund their ideas. The work required for a successful crowdfunding campaign is comparable to that involved in finding a venture capitalist or angel investor to back your business or idea. Do not let the success stories deceive you into underestimating the effort needed for you to succeed.

The Open Source Movement

As stated before, ever since the Internet catapulted crowdsourcing into the public's awareness, many variations of this idea have arisen. We saw earlier that the concept involves many contributors providing time, effort or money to realize a project. The first example, crowdfunding, is revolutionizing the way new products and services are funded. The second type of crowdsourcing, the open source software movement, is transforming the way we write computer programs.

The fundamental importance of open source software should not be undervalued. In raw numbers, the new way of sourcing code is estimated to save consumers up to $60 billion per year in the software arena alone. This is enough to hand out a free laptop to every person in the entire country of Mexico, or twice what is needed to feed the world's undernourished.

This software paradigm provides the world economy with the necessary innovation grease to keep many smaller and less profitable ideas turning. License-free software allows artists and other hobbyists a way to pursue advanced projects without large upfront investments. It has branched out from its original focus on writing code to other areas such as biotechnology, robotics, architecture, medicine, and many more.

When people think of the concept of open source, often their belief is that it is software which is free of charge to be used and modified at will. This is not quite the case.

The words "open source" were first used in connection with software, and at a very basic level it was about guaranteeing a certain type of access to anyone wanting to work on something (code, hardware, medicine, etc.). One basically exchanged universal help for the right to have exclusive access to one's own work.

Open source can be likened to open viewing rights—if you alter, improve or change the design or blueprint of a product in any way, then anyone has the right to see what you've done and use or change your modifications. As expected, many variations have since developed on this theme according to what you want other people to be able to see and do with your and the previous contributors' efforts.

This can be compared to what has been called "free" software. As with its related cousin, there are also many variations of the word free. The interpretations span everything from a completely free version of code or software,

which can be used for any purpose without any restrictions, to free trials, or "freemium" versions, which allow restricted use of code or software. This is often
achieved either via restricting functionality or the time
span for which the software can be used. The most radical
definition of free is that given in the Free Software Movement's manifesto, where the users have the freedom to
run, copy, distribute, study, change and improve the software.

Programmers subscribing to the Free Software Movement's definition allow access to their code, but require
anyone using their code do the same. So if you run a company and wish to use such code, then you cannot expect to
be able to keep your adaptations and improvements secret.
You are legally bound to reveal anything you do with the
code to anyone who requests it. This obviously makes the
use of such code difficult for many commercially focused
entities.

The words "open source" were deliberately chosen to
avoid confusion with the free software promoted by the
Free Software Movement. The term became well-known
due to a daring move made by the developers of the
Netscape web browser. In 1998 Netscape decided to give
away the source code to its flagship product, Netscape
Navigator. This action sparked a number of other developments leading to a famous summit held in April 1998,
called the Freeware Summit. The press release distributed
after the meeting put "open source" software squarely in
every programmers' focus. The term was created as a
counterbalance to the Free Software Movement's goals,

which were seen as too commercially unfriendly and re-strictive of code usage.

This was not the first time that people had sought to set a new agenda of publishing commercially friendly code. The original MIT software license predates the coining of the open source word by about a decade and is one of the most popular and liberal programming licenses available. Companies around the world use MIT licensed code in thousands of products every day.

Although the term "open source" is recent, the idea of open source has appeared in numerous, more or less re-strictive, forms throughout human history. For example, guilds and associations in the past granted their members free and unrestricted access to other members' inventions and ideas, but those outside such closed door associations were locked out.

With the arrival of the Internet, open source was able to take on a different meaning. Instead of sharing within organizations or geographic regions, the information could be shared with anyone having an Internet connec-tion. Admittedly, this does not include those without ac-cess to the Internet, but it presents for us the most realistic possibility of openly sharing knowledge with almost every-one on the planet.

The current open source movement is implemented via an agreement, or license, with users of the shared knowledge resource. With code, it is an agreement which specifies how other users are allowed access to the pro-gram after you modify it. In its most restrictive form, any-one who changes the code must grant anyone else the rights to study, change, and distribute the altered software whenever they like e.g. GNU GPL. In its least restrictive form, modifiers of the code merely have to give an attribution to the original writer of the code e.g. MIT License.

This difference is often most relevant for those wanting to use code for commercial purposes. The latter means it can be used by any business without having to show what changes they have made to the original code. The former means that all changes and additions to the code conducted by a business must be made publicly available.

As with all large collaborative efforts, there are many differing perspectives and opinions, and one of the core licensing challenges has been to try and unify them. This is now reflected in the number of possible licenses available in the open source sphere. There are more than 1400 options. Despite the license proliferation, there are a few well-known and standard licenses such as the aforementioned MIT License which have paved the way for industrial usage.

If the goal of many of the new licenses has been to open up code to commercial usage, unfortunately, instead of simplifying commercial exploitation of software, the license overgrowth has created a legal quagmire bogging down companies trying to use any freely available code. Firms often require legal departments to approve the use of the particular license for open source code before its employees can take advantage of the free and publicly available software. This would seem to be in contradiction with the open source movement's stated goals.

If anyone from the Free Software Movement (which has more restrictive conditions on code sharing and sees open source as "amoral", according to its website) had wanted to stop the open source movement, then there seems to be no better way than creating endless license variations and promoting them. The many license options just help obfuscate the legal issues surrounding code sharing and make companies less likely to share and incorporate external sources of code.

You may ask, what would happen if you just ignored a software license and did what you wanted with it? Can someone come after you to protect the interests of some programmer sitting in Azerbaijan, writing an esoteric browser plug-in?

In 2008, the U.S. Federal Courts of Appeals confirmed that free software licenses create legally binding conditions on the use of copyrighted software. Hence they are enforceable under current copyright law. If you ignore the license agreement, your license terminates and you are infringing copyright.

So what is open source's Achilles heel? A complaint often made against open source software is the quality of the code. In general, this only seems to be a real issue for smaller projects, although larger projects such as SourceXchange and Eazel have also failed. Generally speaking, larger projects which attract hundreds or thousands of contributors seem to function quite well, with rules and guidelines defining how changes can be made and by whom. The level of management is such that, in some projects, minor changes can demand the attention of two independent programmers. This is better quality control than in many for-profit companies.

Another objection raised against open source is that it allows hackers knowledge of weaknesses or loopholes in the software which may be harder for them to find in closed source software. This may be the case, but it also means a contributor can find the same problem faster as well. In open source projects, there seems to be adequate response to discovered issues, and in fact, it seems to be much better than the response by some multinational companies to their known security flaws.

The Big Open Source Projects

On the whole, the open source software scene is thriving. Big, well-known open source products such as the GIMP image editing system, the MySQL database system, and the LibreOffice office productivity suite, are currently the go-to products in their markets. However, the biggest and most well-known open source software is installed on what most people carry in their pockets. At the time of writing it was the most used operating system in the world, with around one-sixth of the world's population using it.

The platform is owned by one of the Internet era's behemoths. It is Android, and together with its owner, Google, it defines hundreds of millions of people's Internet experience. In spite of the Android operating system's world dominance, its birth was anything but smooth. In fact, the company which developed the code base, Android Inc., hit the wall in 2004 and only survived because a

$10,000 injection of funds allowed the company to stay afloat.

Not only did Android have money problems, it also had product issues. The original purpose of the Android team's software was to serve as an operating system for digital cameras. However, given the meager economic opportunities presented by the market for digital camera operating systems, the company pivoted and attacked the smartphone operating system market instead. This decision proved to be a good one because within one year, Google came knocking on their door with a check for at least $50 million to buy the company.

From this point, Google developed the operating system further until in 2007 it was revealed, via the Open Handset Alliance organization, that Android would be the Alliance's first software product. It was to be an open source smartphone operating system built on the Linux kernel. One year later, the first smartphone running the operating system, the HTC Dream, was released.

Android has grown to be the smartphone's most popular operating system. In 2013, over 80% of all smartphones shipped around the world had a version of Android on them. But its dominance has also spread to tablets and is soon (or has already begun) to enter all sorts of connected devices, such as cars and TVs. And what does Google gain from this? Their search engine and related software is pre-installed on billions of devices.

Android has been a massive financial success for Google, but what about the other open source projects? Although open source means free, it doesn't mean that all open source developments are unprofitable for their creators. Many companies are building or have built their business models on open source software.

Such business models are interesting because when compared to the much simpler closed source way of doing business, where users or customers pay for software, it is not always clear how developers can make money when they open-source their main product.

As improbable as it seems, a number of companies have become quite profitable from open source software. Some of them are even multimillion-dollar organizations. Although it is not a major money maker for the company, even Apple offers an open source version of its operating system, called Darwin, for people not willing to pay for the standard version.

Probably the best-known example of an open source product, which is also a profitable business, is the Mozilla range of products. In particular the Mozilla Firefox web browser.

At the time of writing, Mozilla Firefox is the second most used desktop web browser, beaten only by Google's Chrome browser. It runs a successful business model whereby other Internet giants, such as Google, pay to have their webpage displayed on the opening page or installed as a default search engine.

Firefox has an interesting and innovative history in its own right. The interface to the Internet has its roots in the once dominant Netscape Navigator. Netscape Navigator was one of the first web browsers and at one point had almost 80% of web traffic accessing the Internet via its software, at which point Microsoft decided to declare war. From that point the usage of Netscape's software went into decline until the company was eventually bought up by the media giant AOL.

Despite the demise, the browser project never really died because its code was released under an open source license in 1998. This gave birth to the Mozilla suite of products which eventually led to the Firefox web browser which can be considered as a type of successor to Netscape Navigator. Because the software is open source, it benefits from hundreds of contributors adding new functions and capabilities to improve the Internet browsing experience for all users.

Firefox is a great example demonstrating the importance of risk sharing amongst the project's contributors by allowing thousands of people to make small contributions. Numerous open source software programs are leaders in their product classes because they enable people to make improvements to the code base easily. This way of working shows how lowering the risks for an individual can drive enormously innovative and socially important outcomes.

Other Open Source Models

Outside of the coding arena, many other open source models focusing on other things have been quite successful, reaped mass media attention, or both.

Numerous projects in the area of robotics and hardware related fields best demonstrate the success of such activities. An example is the ubiquitous, at least amongst hackers, minicomputer board named Raspberry Pi. Because it provides a basic circuit with several useful hardware items, it has been used to design smartphones, game consoles, radios, robots and much more. The Pi's design is open and despite this, has been an enormous success. At the time of writing, it had sold over five million units around the world.

And although very successful, still, the Raspberry Pi is not alone in the field of minicomputers for hobby computing. There are other successful open source microcontrollers such as the Arduino, which address different user needs (e.g. precise clock cycle timings for 3D printers). Many smaller niche providers have arisen to appeal to all segments of the market. Even the BBC has joined their ranks with its own tiny computer, the Micro Bit.

The open source nature of such hardware has pushed the field into prominence, in part, due to its accessible design which allows people to really understand and tinker with the basics of the circuits. It may seem paradoxical that Raspberry Pi or Arduino could succeed because an open design allows other companies to easily copy the designs. But as a community gathers around such devices and creates online resources and help forums, this makes it harder for lookalikes to break into the market.

When it comes to attracting media attention, few projects have done better than the ones started by Defense Distributed. Their stated goal is to allow anyone to create "a working plastic gun that could be downloaded and re-produced by anybody with a 3D printer." The design is completely open and has caused a verbal shoot-out between both firearm supporters and opponents.

The project has been described as everything from a public safety threat to Armageddon. The reaction to a video in which the Defense Distributed founder fired the first 3D printed gun was swift. In response, a New York congressman announced that he would introduce legislation to ban making such weapons. The cyber weapon producer's retort? "Good F&#%ing Luck."

The first round of home-made gun designs were 3D printable for anyone with the suitable hardware. Many professional weapons specialists questioned the reliability of such parts. However, with their latest foray into this field of "wiki weapons," Defense Distributed is attempting to bring the production of high-powered weapons into the hands of genuine amateurs.

One of their most recent fireworks, the "Ghost Gunner," provides buyers with a design for a simple CNC mill to produce the legal crux of an AR-15 semiautomatic rifle, the lower receiver. This is the part of the gun holding together the stock, grip, trigger assembly, magazine well, and most importantly, the manufacturer's serial number.

Whether this weapon is considered an "assault rifle," and therefore subject to regulations in some U.S. states, is determined just by this piece. It is now possible for this firearm to be manufactured at home with the open source tool from Defense Distributed. The new "Ghost Gunner" mill makes it possible for anyone anywhere to produce a powerful, unregistered and untraceable weapon.

Finally, open source biotechnology is an area that seems to demonstrate early successes and is also attracting media attention. Biotechnology seems the perfect receptacle to benefit from an open source model—if everyone can agree on the details. Such an agreement would be the perfect fuel for the garage-based biotechnology entrepreneur which could drive the next startup revolution in the life sciences. One only needs to look at the software world to see what is possible.

Startups already have begun developing the next wave of biological tools, designing organic systems formulated on basic "wetware." One version of such wetware comprises a set of DNA building blocks that can be put together to produce organisms with a pre-programmed functionality. Imagine taking the DNA of a bacteria, adding one of the building blocks such as DNA from a glowing jellyfish, and the result would be a new glowing organism.

The BioBricks Foundation, a non-profit organization, has already begun championing the fabrication of standard biological building blocks or "biobricks" which, when assembled, should engender functioning synthetic biological systems. These biological systems will be tailored by their creators to carry out specific functions, determined by the biobricks employed to make the synthetic organisms.

In fact, it was the synthetic biologists who led some of the early work on innovative licensing agreements in biology. Their major goal is to produce new DNA based life forms. Achieving this is only possible with cheap and easily accessible research and components. The tool to realize this is the open source model, and the ardent proponent who is trying to make this a reality is entrepreneur and biologist, John Schloendorn.

Born in Germany, John has always shown an aptitude for high quality science and business. He has served as the CEO and director of a number of biotech companies and research centers, with his latest being Gene and Cell Technologies. The native German became passionate about the open source biotechnology movement and started advocating the new paradigm's advantages over services provided by closed-door companies delivering overpriced but cheap-to-produce products.

An unassuming vial with a barely visible speck of goo in the bottom would not seem to be too valuable. However, in the world of biotechnology, it's worth thousands of dollars. Milligram samples of certain types of proteins can cost hobbyists and university laboratories alike a small fortune. What are these expensive ingredients? Highly refined sugar and salt water. The products cost almost nothing and can easily be made by an undergraduate student. Given how cheaply large volumes are manufactured, the prices simply cannot be justified, even if initial capital costs are included.

It is this blatant profit mongering, or more politely put these barriers to entry, which galvanized John Schloendorn to throw his efforts behind an attempt to pry open the field of biology to make it more accessible, and thus allow more progress to be made in biotechnology. The afore-mentioned prices for the necessary research components are only one hurdle, however, the agreements purchasers of such products are made to sign is the other major obstacle. They prevent scientists making their own generic copies of the proteins and selling them.

The market for such products is restrictively protected by a small set of companies hoarding large profits, using the intimidation of complicated legal barriers. This exploitative edifice is what the young German hopes to tear down by open-sourcing the necessary designs and products. It is a work in progress but one that is moving forward quickly.

Companies like Gene and Cell Technologies are becoming focal points in a rapidly changing industry. They promise to "prick the bubble" of high prices and remove unnecessary restrictions on biological reagents and their use. If successful, they could usher in a new period of innovation in biotechnology similar to that seen on the software front. And if it is achieved, it will allow faster and cheaper experimentation which could lead to a genuine revolution benefiting our health and the health of other life in the world around us.

The Point

The many forms of crowdsourced assistance presented here have demonstrated an important way for innovators to reduce their exposure to risk. They spanned the spectrum of asking the public for their time to solve a particular problem (and perhaps in the meantime also raising public awareness), to asking them for their money and thereby sharing the financial risk of a new undertaking.

How many people do you know who have said they could not start a venture because they didn't have the time or money? Now this excuse largely no longer exists. Apart from providing labor and finance, the helpful masses also offer a third advantage for innovation woes—gauging the market. If you think you have a fantastic idea, then there seems to be no better, low-risk way to test this than by crowdfunding your innovation.

In practice, of course, there are still real risks associated with running a crowdsourcing or crowdfunding campaign. Fundraising is not completely free and many successful campaigns have probably only been successful because they invested a lot of time and money into creating a marketing campaign to wow potential donors. Fancy videos, graphics and media presence can turn a lackluster campaign into a viral product infecting the hearts and minds of millions.

In addition, you have to be realistic about the medium that is being used. Tickling the fancy of a very large group of people can be a major challenge. Numerous writers have complained about the whim of the crowd, but if your project has real mass appeal, it may just be worth the effort.

The real enabling tool, which can take something from obscurity to the center of the world's virtual stage, is the Internet. By simply increasing a campaign's visibility, achieving highly funded products such as the Pebble are now possible. However, we are only beginning to scratch the surface of what is possible using the Internet to enable projects to reach very large numbers of people.

Democracy has always attempted to allow the "crowd" or the people of a political state to decide who their future leadership will be and what its direction will be. The reality, however, has been a continual process of weakening and eroding this ideal until people feel their welfare is largely disregarded and there is less and less they can do about it. This may be about to change.

Current Western democratic systems tend to share the following features. Once every few years, people are given the chance to elect the decision makers for their city, state or country. In the meantime, the politicians chosen by the majority decide what happens and what does not. So if the elected representatives do not fulfil their election promises and just, in effect, pursue their own interests while in power, then there is often little recourse available to the public to prevent them from making bad decisions and behaving this way, until, theoretically, they receive judgment on the next election day. The Internet though, is offering a way to change this.

With the emergence of the connected world where information is only a click away, people can now get involved in any issue, no matter how minor. One can argue the public now has the means to set the national agenda and help the government make decisions the people desire. In fact, it seems that the Internet may allow us to completely change the way democracy works. Instead of electing people to govern and decide on every matter, these people will simply become figureheads. Each time an important decision needs to be made, the responsible politician could be required by law to turn to the crowd, since the elected official's job is now just to carry out the crowdsourced decision. The populace will never again have to complain about the people they elected!

This of course is just theory and conjecture, and many practical difficulties would need to be considered, but it has been raised as a way to overcome the partisan gridlock that so often cripples democracies. Some countries such as Switzerland already have an approach resembling this, even before the Internet simplified its implementation. The small, landlocked nation allows its citizens to challenge and vote on all sorts of laws via referendums and initiatives if they can garner enough support.

Such a powerful system of governance has the potential to solve many dilemmas caused by modern day politics. Under this model, politicians would no longer have the power and responsibility to run the nation on their own. They would simply have the mandate to execute what the public tells them for each decision. Practically, this could mean all small decisions are made by the politicians, but decisions affecting large parts of the community or ones above a certain monetary value, are made by the population at large. If voting is non-compulsory, it has been suggested that only those affected by particular issues would bother to vote on them, and it would not be a burden on all of society to vote every time.

For politicians, the benefits could also be significant. It would greatly reduce the need for constant opinion polls. The government could not be responsible for making decisions contrary to the will of the majority, because each time they would only be implementing matters agreed on by the citizenry. Certain issues of corruption and influence would also disappear as politicians become rather like project managers making sure the public's wishes are carried out. What an idea!

Given the praise this chapter has had for the crowd, there is still one major question not covered. Are crowd-solved problems any good? World-renowned network and social scientist Duncan Watts in his book, *Everything is Obvious*, demonstrated several cases where the crowd did not perform as well as hoped. His research arose from the desire to know how well the masses could predict the outcomes of certain events.

To examine the intelligence of the masses the researchers took advantage of a powerful information aggregating tool known as a prediction market. Although not new, interest in them has reignited, especially due to books like the New York Times bestselling, *Wisdom of the Crowds*, which popularized their effectiveness. These are markets where participants can wager on the probability of an outcome of an event. Market prices are set according to what the public thinks the probability of an event is. In other words, what people are willing to pay given their perception of the current chance of the outcome happening.

For example, one may have the opportunity to buy a contract which pays $1 if a given horse wins a horse race. It is up to a buyer to determine how much such a contract is worth to him or her. If he or she thinks the horse has a

40% chance of winning, then the bettor will not pay more than 40 cents for the contract. And so through this mechanism and observing the current market price, we can determine what the crowd thinks the probability of a certain outcome occurring is. It is important to note that what we are trading here is information and not bets on a race winning horse given a bookie's predetermined odds. The bettor chooses his or her odds and its price. We are betting on a given outcome occurring, whether that is a horse winning a race, a president winning an election or a spaceship successfully launching into space.

It can be quite time consuming to set up mechanisms to gather and filter crowd-based suggestions in prediction markets, so the researchers hoped the effort would be justified and their research would show the crowd could make better predictions than a simple statistical model. Unfortunately, this was not the case. Predicting both the outcomes of Major League Baseball and NFL League games seemed to stumble the crowd as its performance was indistinguishable from a simple statistical model.

Although there exist situations where prediction markets outperform expert opinions and other well-informed information sources, this final example reinforces several points mentioned earlier. The utilization of the crowd in this way is a fairly new phenomenon whose nature and limitations are yet to be fully explored, much less understood. And it is always important to understand the medium being employed, and not to expect it to be suitable for every goal.

When using the crowd to help us, we need to handle it with care. Whether for politics, prediction or play, carefully designing methods to involve the masses is critical to success. If we mess that up, then as the old software saying goes, "garbage in, garbage out."

And, unless you're Thomas Edison, isn't innovating with garbage rather difficult?

Chapter 3
The Latest Tools And Space—For A Dime

Once known as the Kulin nation and home to the Wurundjeri, Boonwurrung, Taungurong, Dja Wurrung and Wathaurung people of aboriginal lineage, the city of Melbourne has always had a multicultural landscape. Although the original inhabitants have become a minority after the arrival of European settlers in August 1835, the city has maintained its culturally diverse population.

At the beginning of the 21st century, Melbourne was home to more than 116 religious faiths and immigrants from around 180 countries speaking over 233 languages. This mix has led to the city becoming the second most culturally diverse place in the second most multicultural country in the world: Australia. As many locals will proudly declare, it has the largest Greek population outside of Europe after Athens and Thessaloniki.

Although Melbourne's multinational character may not be obvious to everyone, its diversity is infectious. It exhibits the flair of a large cosmopolitan city while at the same time successfully preserving its old world charm. Since the first settlement by the British, it has grown into a bustling, international hotspot with over four million inhabitants.

The city is sprawled out over an enormous area of land with the northwest pocket of Melbourne more than 70km (or 40 miles) from the southeast corner. In the middle is the business district which is the envy of many capitals around the globe. It has world famous beaches less than 15 minutes from the heart of the city which help contribute to it being consistently ranked in the top five most livable cities in the world. Let's explore a bit of this modern metropole.

If you were to walk towards the historic Flinders Railway Station from the city center, you would be greeted by an iconic building. Its magnificent ochre brick façade, facing Melbourne's busiest pedestrian crossing, bears a large Renaissance green copper dome flanked on both sides by two smaller cupolas. At one end of the building, two city blocks away, stands the station's majestic clock tower of 1883. At the other end is the entrance to what was the world's busiest passenger station in the late 1920s.

Ascending the flight of steps under the building's arched entrance, the first feature that catches one's eye is the distinctive nine timetable clocks dating from the 1860s. Given features such as these as well as the historic importance and architectural grandeur of Australia's first railway station, one can understand why locals protested numerous times to it being demolished and replaced by office towers and a shopping mall.

Once inside, you must first pass through the turnstiles and down one of the many escalators leading to the platforms to get to the trains. From there, a 15 minute train ride east will take you through a typical cross section of the modern city's landscape. Passing tightly packed houses and buildings, interspersed with lush green parks, much of Melbourne's juxtaposition of old world charm and contemporary design can be found at the end of a short train ride in the suburb of Hawthorn.

Home to the world-renowned Swinburne University of Technology, this area is bustling with students during the week. In the afternoon on most weekdays, a small group of another kind of learners meets for a session of tinkering and playing with hardware in one of the back alleys a stone's throw from the University. In a small garage with yellow Plexiglas doors and an inviting LED cable strung across the doorway, you will find the Connected Community Hackerspace Melbourne at 5 Kent Lane.

The hackerspace's appearance is true to its roots. It has a welcoming workspace extending about 20 meters into a cozy back area where people can relax and talk about their latest project. The entrance sports a hacked soft drink fridge—a staple of most hackerspaces.

The entire hackerspace is a smorgasbord for DIY enthusiasts. Tools and electronics line the shelves, with half completed "masterpieces" strewn between them. There are colorful spools of plastic cords against the walls, resembling reels of multi-colored licorice strings in a confectionery store.

In the back corner are organized drawers of LEDs, resistors, capacitors, wires and other electronics used as essential building blocks for all types of wacky projects. Beside the electrical treasure chest are many devices for testing and building, such as soldering equipment and an oscilloscope for electronic experimentation.

Still glancing down the end and taking a deep whiff, one notices the whole area smells of acetone, which is used to provide 3D printed objects with a smooth and professional finish. Looking back, the ground has the usual splatter of color and plastic coverings indicative of an unfinished room, giving it the feeling of a work in progress—just like most of the objects located on the bench tops.

It is Tuesday night and there is a buzz among the hackers. True to hackerspace tradition, on this evening the space is open to the public—people can come in and experience the hacker life for a night. A dozen new prospective members are standing around while veteran hackers chat convivially about their projects.

Projects range from handcrafted, half-built drones with their rotors and wiring exposed (the frame of which was printed on the 3D printer standing against the wall) to a 3D printer that was invented in this hackerspace. At present, the home-made printer is slowly rendering a rectangular scaffold. Busily it prints, with submillimeter accuracy, a large black square, which takes shape as something vaguely recognizable, possibly a frame for another invention. Its mysterious sculpture will only become clear after 12 hours of meticulous printing.

Rob, a regular to the hackerspace, is the proud creator of the 3D printer. He is a focused man who looks to be in his late thirties. Unshaven and fiddling intently with some wires, he is working on improvements to his new 3D printer. This year he built it from scratch using scrap machines that were lying around the hackerspace.

On a shelf, relegated to the secrecy of the back corner of the building, sits a RepRap 3D printing device waiting quietly to replicate itself. Its place here is part of the ethos of creating more than just trinkets and plastic toys. The dream of creating a machine that can replicate itself appeals to the ideals of those who work here—whether that be for serious fun or to realize something more meaningful.

The Connected Community Hackerspace Melbourne was founded in 2009 by Andy Gelme. An ebullient individual with graying hair and a sleek leather jacket on this evening, he appears to be in his late fifties. When engaged in conversation, he quickly rattles off hacker related facts in a monotone voice and is elated to tell you about his prized possession—a large computer container.

Andy's claim to fame is his relationship with Cray. Cray is one of the oldest supercomputer manufacturers in the world and its machines are famous to any computing devotee. In the backroom of the hackerspace headquarters is the first Cray X-MP computer chassis that arrived in Australian in 1986. This antique piece of hardware is a testament to the hackerspace's old school feel and embodies the venue's cluttered give-it-a-go atmosphere.

Not only does Andy have the chassis for the first Cray in Australia, he was, in 2010, also the only person in the world known to still own an operating system to run this particular computer. So when Chris Fenton and Andras Tantos, two electrical engineers, wanted to reboot their

Cray on the other side of the world, it turned out Andy still had a disk pack containing an original copy of the Cray operating system. He had been waiting to put it to use for over 25 years.

This all eventuated when Fenton reported his work on the Internet and the publicity it generated caught Gelme's attention. At once, he could see a wonderful opportunity for the old software and found a way to get it unscathed to New York. When the disk arrived, Tantos took responsibility for working on it. He rewrote the software to provide support for hardware devices including printers, monitors and keyboards amongst others. It took him nearly a year to complete but now the Cray works.

This type of achievement is highly illustrative of the hackerspace's do-it-yourself attitude. It is an infectious collaborative approach and it is why hackerspaces have been so important for startups and innovators around the world.

A critical part of innovation is creating your first product or service. Whether you are creating software, hardware or simply an idea to be sold to the world on a website, the development of the first prototype is an essential part of keeping risk under control—it is never a good idea to bet everything on an untested concept.

It certainly would not be very helpful if the creation of your first prototype led to bankruptcy. This is why, for many innovative ideas, it is critical to always start with some type of a Minimal Viable Product, i.e. the low risk proof of concept for your customers.

For startups this is logical. As it is common of most fledgling organizations to be strapped for cash, it pays to create a bare-bones model as cheaply as possible. For larger businesses launching hundreds of new product lines, this idea is also a good one and for the same reason. However, what if producing your first prototype requires several expensive machines? What if you cannot produce it without dishing out $100,000 on a computer controlled milling machine? The answer: visit your local hackerspace.

A hackerspace is typically a community run workspace where people interested in learning and building things can meet. It is a place to share knowledge and experiences, and to "hack" things. Here objects like robots and circuits are made to carry out tasks not originally intended for them.

Because the distinction is often lost, one regularly hears people referring to hackerspaces as "makerspaces." Though, on the whole, this would seem to capture how these community workspaces see themselves, for some there is a difference between the two words. Hackerspaces are claimed to be typically about bending the original purpose of hardware and software to produce unintended outcomes. Makerspaces tend to be more about making things with off-the-shelf products.

Besides these two styles of community workspaces, many others have joined the fray. TechShops, FabLabs and others also provide access to tools and software, as well as training and education. Some of them are for-profit franchises and others are simply based around a strict set of rules that usually require users to donate money toward maintaining the space. Their goals, however, are similar—empowering people to build their dreams.

Hackerspaces were originally created for like-minded people to meet and exchange ideas. They eventually evolved into a common space where people could cheaply rent or use machines, electronics, and software that would normally be unavailable to average users. And most remarkably, this movement all began with a small group of like-minded computer enthusiasts in Germany who would become the inspiration for thousands of others around the world.

Hackerspace Genesis: The Chaos Computer Club

The seed for the Chaos Computer Club was planted in Hamburg on September 12, 1981, when five German friends met to talk about technology and other related issues. In contrast to prevailing public opinion that feared computers would only bring about more surveillance and fascism, this inaugural group of young hackers had the foresight to see that socially beneficial outcomes could be achieved with new computer technology. They called themselves the Chaos Computer Club.

The Club gathered to talk about cryptography, digital bulletin boards and amateur radio. They also built computers. True to the subversive spirit of hacker groups then and now, they began their activities hiding in the shadow of the German telecommunications giant: Deutsche Telekom. The large firm had a monopoly on telecommunications, and in the eighties it was a crime to connect anything besides a telephone to the telephone network. This meant that if you were caught hooking up a modem to the network without an official seal, you could go to jail for up to five years. By connecting their modems to the telephone network and exploring the digital world, the group was risking serious prison sentences.

The group's first adventure occurred in the late eighties when a petty criminal, with a loose association to the hacker club, contacted the KGB. He offered them a list of American computers he had hacked. Driven by the need to finance his cocaine habit, he had hacked into the Advanced Research Projects Agency Network (ARPANET)—the forerunner to the current Internet in America—with the hopes to sell the results of his work.

ARPANET was a project funded by the United States Department of Defense to disrupt the way we communicate. Instead of dedicating a copper wire between two people for communication or data flows, part of the plan was to divide the data into packets and send it around a network. This allowed a single connection to be used by multiple people and paved the way for the modern internet.

Although the initial goal of ARPANET was, according to the former Defense Advanced Projects Agency Director Stephen Lukasik, "to exploit new computer technologies to meet the needs of military command and control against nuclear threats," back in the early 1980s, ARPANET didn't have missile launch codes or other confidential information. Despite this, it still contained interesting facts about the United States' infrastructure.

So when the story that a member of the Chaos Computer Club had hacked into the new American communications network hit the press, it showered the club with unwanted attention. More frightening for the hackers was the fact that shortly after the incident came to light, the addict was found dead under suspicious circumstances.

To make matters worse, public opinion on hackers had also shifted. In the media, the word hacker had become tarnished. Previously, the term had merely referred to people "hacking" together hardware or software and had not been perceived in a negative way. Hackers were now seen as spies and criminals.

Fast forward thirty years and today's club is spread all over Germany, but its head office is in Berlin. The current headquarters has been in Marianstrasse 11 (an old carpenter's workshop) for nearly twenty years. The move to this address two decades ago was one of many turning points for the association—it granted the club a permanent internet connection that attracted many new members.

Over the more than thirty years of existence, the club has explored many aspects of computing and security flaws have always been of particular interest. One of their earliest high profile demonstrations was in 1984 when they committed the first electronic bank robbery by hacking a German videotext machine. They used it to transfer money to their bank account. After turning themselves in as a public demonstration to show that security concerns were serious, they have continued to expose other high profile security weaknesses.

They staged another spectacular display in 2014 when they hacked biometric data from the former German Defense Minister, Ursula von der Leyen. They were able to recreate her fingerprints from publicly available photos and, at the same time, made evident the loopholes in fingerprint security systems.

Clubs such as the Chaos Computer Club have given rise to communities with a similar mindset looking to share knowledge and tools. The members hope to achieve more than they could on their own. These ideals and motivations have formed the basis for many similar hackerspaces around the world.

Since the first meeting of the Chaos Computer Club in 1981, hackerspaces have been founded in almost every major city worldwide. As of 2012, there are an estimated 700 to 1,100 active such community workspaces all over the world, and they are multiplying. The common characteristic of all these communities is that for typically less than $100 a month (at the time of writing), they give members practically unlimited access to a workplace with powerful equipment and tools, and just as importantly, potential stimulation and valuable support and collaboration with similarly minded people.

The Modern Hackerspace

The most important and often overlooked part of a hackerspace is the building it occupies. The correct DIY edifice is vital because it provides the physical infrastructure that members need to work on their projects and collaborate with peers. Besides workbenches and the like, most hackerspaces offer electrical power, computer servers, and a network with Internet connectivity.

Walking through a typical well-equipped hackerspace will expose you to a cacophony of smells, noises and machines, including the usual sound of drills, soldering equipment and general banging. Like a manufacturing site with a relaxed experimental atmosphere, wandering around gives the impression of being in a haven of innovation.

If we were to start a tour of such a building and begin at the back of any given premises, we would normally see some of the more dangerous equipment first. This is why it is located here. One particular machine would seem strange as it whirs and moves around randomly above a wooden or metal object. Intermittently it shoots a laser into the object and leaves burn marks with no apparent meaning or purpose in its choice as to where to go next. The contraption in front of you is a laser cutter, and moreover, is highly dangerous, sporting a beam powerful enough to blind you instantly or worse.

Although laser cutters have become ubiquitous, it is still quite common to see other cutting equipment such as plasma cutters, water jet cutters and knife cutters. Like the laser cutter, most of these are also computer controlled. Such machines can be used to cut and engrave all types of materials such as wood, plastic and sheet metal. Some cutters function by producing a jet of energy that burns into a surface. Others slice with the force of water or a blade. A cheap laser cutter with software costs a few thousand dollars. A good one will set you back tens of thousands of dollars. It is clearly a bargain if you are only paying a hundred dollars a month to use these machines—not to mention access to everything else.

As we walk back to the entrance, somewhere along the way we are almost guaranteed to see a machine chattering about with a robotic whirring noise. If you look closely, you will see it move a spinning drill bit above a pile of metal shavings strewn about a block of deformed metal. This machine carries out its motions relentlessly, focused on carving away metal while being phlegmatically guided by code directing it from a nearby computer. The computer controlled mill frequently rings out high pitched screeching sounds from the bore head as holes are precisely drilled by a fast-spinning cutting tool shaped like an L but with a short base.

Mills can sometimes be accompanied by another machine called a lathe. It has a similar purpose, removing material to transform a nondescript block of metal, plastic or wood into a useful component. Instead of having a drill poking out and cutting away material, a lathe tends to have a large tapered cone spinning at high speeds to grind away unwanted parts. A cheap mill or lathe can cost roughly $5,000—$20,000. A high quality machine with submillimeter accuracy will set you back up to fifty grand.

To run this equipment, computers controlling the machines need data to tell them what to cut and when. This requires software to design the final desired object, which likewise has a price tag. Luckily, this technology is also usually a part of what is offered in hackerspaces.

Hackerspaces often have computers with programs able to visualize accurate pictures of many useful parts ready to be made. Typically with the abbreviation CAD (Computer-Aided Design) in the title, this software is essential for designing high quality duplicate parts. Over the years the costs of such software has dropped dramatically, and larger companies such as Autodesk are even offering free software for basic design activities.

The few machines we have looked at so far and the software used to run them merely scratch the surface of what can be available. Larger hackerspaces can stock more exotic items such as 3D scanners, heat strip bending systems and much more. However, one piece of truly iconic hackerspace equipment is found in nearly every collection, sitting in a corner slowly buzzing away for hours with little supervision. Its whirring movements may look futuristic, but the noises remind one of a standard ink printer. This machine can in fact be considered the printer's successor.

The rapid prototyper, or 3D printer, is the 3D version of an ink printer. In its most common form, it allows a user to create a three dimensional object by extruding melted plastic from a small nozzle that systematically constructs the shape layer by layer. A cheap, inaccurate version of this modern printer costs typically a few hundred dollars. A high quality rapid prototyper, which prints to submillimeter accuracy, can easily empty a bank account of ten thousand dollars. That is why access to a good quality one for a membership fee of less than a hundred dollars a month is a huge bargain.

The wave of media hype surrounding 3D printers has surpassed that of almost any other recent technology found in a hackerspace—or almost anywhere. Perpetual printing, organ printing, and even grander headlines appear in newspapers every day. The key to eternal life? Just print it.

3D Printers

Most people who have heard of but not seen these printers, picture a rapid prototyper as a device slowly extruding melted plastic to build up an object layer by layer. In essence, this image is correct. However, such a mental model only captures a fraction of what is actually happening today in the area dubbed by laymen as 3D printing. In order to obtain some understanding of what can be assembled by these printers, it is worth having a look at 3D printing's humble beginnings and growth.

It is generally agreed that the roots of plastic, consumer 3D printing trace back to the 1980s when new fabricating methods were invented that worked by hardening plastic in layers. Several patents were granted to the inventors in the late 1980s, including one to S. Scott Crump at Stratasys, a company he also founded. In 1989, the young scientist and entrepreneur then received a critical patent for an "Apparatus and Method for creating three-dimension objects."

It took 20 years for this patent to expire but when it did, it created an explosion of innovation. Since 2009, thousands of cheaper and more powerful printers have been produced concomitantly with the rise in crowdfunding platforms. In fact, over 34,000 machines were crowdfunded over the course of four years, making these printers one of the most popular classes of mass sponsored products.

The current method of 3D printing that most people would encounter occurs via the method invented at Stratasys. Among the many descriptive names it goes by, a common one is "fused deposition modeling." In layman's terms, hot liquid plastic is extruded from a nozzle, which then quickly hardens as it comes into contact with the existing material underneath the printer head.

In order to produce a particular object, it is first necessary to tell the 3D printer what to print. This is typically done by creating a virtual, 3D model of the object on a computer in one of numerous possible file formats, of which the most common is called stereolithography or STL. The file content is described by a standard that defines the information to be stored and articulates what the object to be printed looks like in 3D.

Given such a file, it is then possible to open it in specific programs and use them to instruct the printer to do its job. This process converts the computer data into 3D printing friendly code known as G-code, which the printer uses to create the desired object.

At this point you can now print your object. From replacing broken parts on consumer products to specialty toys the possible applications seem endless. Either scan or create your object and you are away.

There are many more types of 3D printing besides fused deposition modeling, and many of these other methods are not just interesting for ordinary consumers but also for industry, to which we now turn our attention.

Industrial 3D Printing

Given that many components used in industrial manufacturing appear to have higher standards than components for consumer use, it may seem intuitive to assume that industrial 3D printing developed after consumer 3D printing became available. Nothing could be farther from reality.

By the 1920s, adding layers of metal as a manufacturing process had already been considered by Mr. Baker, an inventor who had patented a process in 1926 described as "the use of an electric arc as a heat source to generate 3D objects depositing molten metal in superimposed layers." Occasionally, processes were patented in the ensuing years such as Otto John Munz's 1951 invention to create a "permanent three-dimensional record in space of three-dimensional phenomena." In 1971, a Japanese employee at Mitsubishi also patented a new printing process. It was even sophisticated enough to give the walls of the printed object function-specific properties.

A patent fundamental to the methods currently used by the aerospace industry to 3D print complex jet engine parts was filed by a Ross F. Housholder in 1979. Simply called "molding process," it combined a laser beam and a metal powder bed to selectively fuse together parts of the powder bed into an object. Methods such as this can be used to create complex honeycomb style lattices making objects strong but lightweight—exactly what is required of airplanes.

On the surface there would seem to be many similarities between the consumer focused, plastic 3D printers and the methods used by industry, such as selective laser sintering. Both build objects by adding layers of material. They just do so in different ways. Another similarity in both fields was the recent expiry of an important patent for industrial 3D printing. Following this event, we should have seen innovation blossoming, such as when the Stratasys patent's term ended. It didn't happen, and this is where the similarities end.

Industrial style 3D printing, also known as additive manufacturing, is a more complex process and often involves laser technology. Unfortunately for innovators, there are numerous patents already in place that tend to stifle new technological innovation. The large, industrial scale 3D printing machines can cost up to a million dollars. Industry level tolerances and quality demands are a requirement that acts as a further hurdle to small innovators who may want to produce complicated machine designs. As a result, we are not witnessing the same boom in innovative products at the industrial level as in consumer 3D printing. Industry is still hoping it will come, however.

For over a decade, many industries have been investigating additive manufacturing as a means of testing designs, or producing small volumes or one-off prototypes more efficiently. Each industry has its own particular goal, but the one sector that firmly embraced additive manufacturing is the aerospace industry.

The advantages for the aerospace industry are tremendous. Improving basic things, like being able to build engines with fewer parts, or even fundamentally new wing designs for unprecedented aerodynamic efficiency, are now possible. Engineers' imaginations can run wild. However, one of the biggest gains has been in weight reduction and its subsequent fuel savings.

In 2012, General Electric Aviation demonstrated with a prototype that it was possible to reduce the weight of a six-ton gas turbine by almost ten percent with use of this new manufacturing method. The company expects to implement this technology to achieve such weight reductions over the next couple of decades. The net savings would be tens of thousands of gallons of fuel per aircraft per year. If adopted by all airlines, this would be the equivalent fuel needed to run all the cars in America for half a year.

These results are not only good for the economic viability of airlines but also for the environment. A true win-win scenario.

Perpetual Printing

Although the laws of thermodynamics prohibit perpetual motion, there is no law of 3D printers that prohibits perpetual printing. The dream of a machine repeatedly making copies of itself seems to be within reach of 3D printers. One of the foremost projects attempting to achieve this lofty goal is the "replicating rapid prototyper," or RepRap.

Begun in 2005, RepRap is the brainchild of Dr Adrian Bowyer, a former senior lecturer in mechanical engineering at the University of Bath in the United Kingdom. The seemingly impossible printer claims to be the only "3D printer that prints itself." Currently this statement is only valid for its plastic parts, but the goal is obviously either to replace the metal parts with plastic ones, or be able to print the metal parts as well. Despite the grand claim being only partly true, at the time, RepRap was truly pioneering work.

The design of the perpetual printer is open source, meaning anyone can download the designs for free and 3D print the plastic parts. This has led to a proliferation of RepRap designs, tweaked to serve their inventor's purpose. Most likely due to its flexibility, but also due to its open nature, versions of this 3D printer have become the most widely used among the global members of the Maker Community. There may be a financial incentive behind this as studies show that owning a RepRap is an economically sound investment.

Owners and contributors to the RepRap project see themselves as creating nothing less than an industrial revolution in distributed manufacturing. The intentions of its designers are summed up well by a statement in the Guardian: "[RepRap] has been called the invention that will bring down global capitalism, start a second industrial revolution and save the environment..."

Printing The Future

If perpetual printing is not groundbreaking enough to impress you, then what about printing a new ear for yourself—a fully functional, anatomically correct body part? Current projects around the world with goals like this are pushing the boundaries of what was thought possible for 3D printing.

3D printing initially focused on more traditional manufacturing but now no industry is excluded. Researchers are racing to perfect methods that could put an end to patients desperately waiting for available organs. Thus interest in making on-demand body parts has skyrocketed in the last decade with dreams of putting an end to the nerve-wracking waiting lists at hospitals.

These breakthroughs have been facilitated by the recent discovery of ways to create human cell scaffolding and to build body parts cell by cell. Future body parts on order seem to be on the horizon. Companies such as Organovo and EnvisionTec have developed devices capable of carefully placing living cells in precisely defined lattices, mimicking the structures of human cells. The work has even gone so far as to show that the cells continue to live and grow after being printed.

It will not be long before we have viable examples of 3D printed ears, livers and blood cells, promising an end to medicine's constant challenge to find organ and blood donors. But the changes are even more impressive than that. Because we control the designs, we can now produce body parts with functionality far superior than their traditional counterparts. Titanium craniums, cartilage or even claws? We can expect these advances to trickle down to benefit the general population over the next decade or two.

Although printing body parts may still seem futuristic, printing our own food is not. 3D printers have broken into the realm of food printing and have started with products such as sugar and chocolate. Producing intricate designs unachievable by human hands has become the norm.

Even more exciting, the companies producing these food-forging machines, such as 3D Systems, are the same companies squeezing plastic out of nozzles next door. They use the same basic concept of extruding a liquid-like substance out of a small nozzle to make food products. So as areas such as consumer 3D printing are rapidly advancing, they are pushing forward the boundary of what is possible in other applications such as printing food.

One of the more common condiments being extruded from a printer head is chocolate. With its malleable properties, it allows artists to create intricate designs. For example, researchers at the University of the West of England successfully printed filigree chocolate designs, raising the standard of what was attainable with common food production techniques.

Mainstream chocolate producers are also getting in on the game. Hershey's has partnered with 3D Systems to allow visitors to design and produce their own chocolate delights at their "Chocolate World" attraction in Hershey, Pennsylvania. This system provides Hershey's with a unique opportunity to discover consumer desires and turn them into new products. There is also the Dutch supermarket, Albert Heijn, which placed a chocolate 3D printer in its bakery department in early 2015. This system was designed to allow shoppers to custom decorate their own cakes.

However, the advantages of 3D printed food are not limited to mindboggling designs. There are also genuine benefits for elderly or incapacitated individuals who have difficulties eating and swallowing. A German company known as Biozoon has developed a so-called "Smoothfood" printer that produces easily swallowed foods catered to individual nutritional needs.

Apart from using the new technology in food production, the 3D printing industry is also tackling another basic human need: personal transportation.

Local Motors, a new motor vehicle company founded in 2007 and focused on low volume car manufacturing, printed the chassis and other major components of a car for the first time at the International Manufacturing Technology Show in 2014. Named the Strati, it took 44 hours to complete. In spite of the headlines it received, the vehicle still required further touch ups and additions, adding three days to its assembly. However, producing a totally new and drivable car from start to finish in five days without a large factory was still impressive.

Local Motors uses a plastic carbon fiber mix to print their car, making them subject to one of the current drawbacks of 3D plastic printers: the speed at which objects can be printed. However, if we take a look at the history of ink printing, which produced around one page a minute in the

1980s, but now can print one page per second, we can anticipate the type of rapid progress that is possible with 3D printing.

In fact, startups and companies have already discovered new ways to massively speed up the plastic 3D printing process, moving the developing of 3D printing technology along at a more accelerated pace when compared with the development of ink printing. Were 3D printers to become as fast as inkjet printers are now, then one could imagine an army of 3D printers replacing the army of humans in factories. For some this may seem like a dystopia, for others a manufacturing revolution.

The diverse applications presented here are a testament to 3D printing's success. It is a success epitomizing the focus of this book—a way of lowering innovation risk. By being able to cheaply test designs, from toys to aerospace, organs to food, the respective industries can now advance much faster and cheaper. Instead of spending hundreds of thousands of dollars to tool up a factory to produce a faulty design, one can find the fault much earlier by creating prototypes. The money and time saved on avoiding poor designs can then be reinvested into other activities and push a company's innovation agenda even further.

How Can I Hack My Way To Success?

Membership to a hackerspace alone is not an instant ticket to success. A good idea and dedication are also prerequisites. Given these two, and a bit of luck, you can build a powerful company at a low cost just like the founders of Square did.

When Jack Dorsey was sitting in a TechShop facility in California in 2008, he was already a prominent figure in the startup world. After co-founding the social media service, Twitter, Jack could have easily retired. But he didn't want to. He continued to hack away at his next idea.

The idea arose when Jim McKelvey, Jack's future business partner, was not able to finalize a $1000 credit card payment. He simply did not have the necessary device to process the payment. This inconvenience gave Jack and Jim the motivation to devise a solution.

Cleverly leveraging the already ubiquitous smartphone and tablet, they exploited their computing capabilities to turn them into portable credit card processing devices. Their first attempts to present the ideas on paper to venture capitalists went nowhere. Without an actual product, they were unable to convince anyone of the merits of their insight in spite of Jack already being a very successful entrepreneur.

They realized they needed to make a prototype. Initially they simply used the camera on a smartphone to take a picture of the credit card. However, there were many problems with this, not the least being dirty cameras and poor lighting. Jim then came up with the idea of creating a simple magnetic strip reader by hijacking the microphone port to collect the person's payment information from the credit card's magnetic strip.

To create their design, they went to a local hackerspace, enrolled in a couple of classes and started prototyping. After they had created a fully functional prototype, they went back to the venture capitalists, confident they could persuade them this time. To show off their new product, they took one of the financier's credit cards, charged $500 to it, and banked their first capital advance.

The company went from one success to the next, and was able to raise $10 million Series A funding based on the convincing demonstration of the prototype. Square has grown to become a billion dollar company with over 400 employees. The company processes around $6 billion in transactions annually. All this grew from a good idea and a $100 a month membership fee at a hackerspace.

Unlike many others, Jack and Jim could afford the development costs of their prototype without a hackerspace. Other developers of products like LIFX, whose humble beginnings were in the Connected Community Hackerspace Melbourne, benefitted greatly from access to inexpensive tools. According to the Kickstarter campaign to launch the new lightbulb, the idea behind LIFX was to create a "WiFi enabled, multi-color, energy efficient LED light bulb that can be controlled with an iPhone or Android."

LIFX may not save the environment or put an end to hunger and poverty but being able to easily and intuitively control a light bulb's color and intensity can be a very useful thing to do. Possible applications range from sleep therapy to disco lighting and security. In fact, the idea for the product arose from co-founder and CEO Phil Bosua's frustration at not being able to turn a light on and off because the switch was inaccessible.

After tossing the idea around with Andy Gelme, the two formed a team and began playing with the idea. They tested several designs and even built a small LIFX box to simplify design tests. This low-cost way of prototyping was made even simpler by also making use of the local hackerspace's tools and equipment.

Phil exploited his marketing knowledge and launched a successful Kickstarter campaign to fund the LIFX project. They ended up with more than one million dollars—ten times what they had originally expected to raise. Since then, LIFX has been able to raise millions in venture capital while successfully competing with manufacturing giants such as Philips. They export worldwide and have moved their headquarters to the startup Mecca, Silicon Valley. However, without an affordable way to test his designs and try out ideas, Phil Bosua may never have succeeded.

Like other users of the Connected Community Hackerspace Melbourne, Bosua and his team benefitted from access to cheap tools to create and test designs. Putting together the circuits without having to buy the necessary soldering and electronics equipment saved them tens of thousands of dollars—money that could be better invested elsewhere in the business, like in a successful crowdfunding campaign.

Maker Faires

If the cramped space of typical hackerspaces is not to your liking but you still want to see what can be made with the tools provided by such community workshops, then you are in luck. A great way to meet and get to know some of the enthusiastic members of such groups is at meet-ups among like-minded hackers and makers. There you can see examples of ingenious projects and applications of some innovative thinking together with the type of tools available, perhaps inspiring the nascent pioneer. The support for these events has been such that they have grown from local get-togethers to major so-called Maker Faires that now take place around the world.

Their humble beginnings were in April 2006, when the Make Magazine organized the inaugural Maker Faire. It was held in San Mateo, California, and was the first in a string of highly successful events gathering people around the Do-It-Yourself hacker and makerspace culture. Since then the fair has grown exponentially and in 2012, it attracted a crowd of 120,000 attendees.

The Chaos Computer Club has also been organizing community gatherings for a number of years, however, the inclusive attitude and breadth of exhibitions at the Maker Faires have set them apart from other such events. They have become so popular that similar fairs are now taking place in the UK, Canada and Africa.

These events are important for creating awareness of and promoting the maker movement. The purported benefits of this empowering mindset are to create more entre-preneurs and allow more people to define their future. It reached a high point in June 2014 when The White House decided to hold its own Maker Faire. This event featured over 100 Makers from more than 25 states and included more than 30 exhibits. The participants were handpicked from the best entries, and some were personally greeted by the President. For many, this was the crowning on their efforts.

Responding to the strong interest generated by hold-ing these events, Make Magazine will even provide help with organizing Mini Maker Faires both locally and over-seas. The demand has been so great that between 2011 and 2013 the number of such events doubled each year.

The Maker Faire events have become a critical part of the maker movement. By providing hackers and makers with social recognition for their toils, it is a great way to strengthen the movement and empower as many people as possible. It shows one just what is possible.

The Point

A critical part of innovation is being able to lower the barriers to make a basic prototype. In the LIFX case, this was accomplished by finding a way to produce a first iter-ation embodying the new idea as cheaply as possible. But hackerspaces do more than simply give inexpensive access to valuable tools—they offer networking opportunities, ac-cess to skills and advice, and a place to rejuvenate when you've exhausted your idea bank.

Makerspaces and hackerspaces are part of a much larger self-empowering movement providing people with cheap (i.e. low risk) access to the tools and know-how needed to design and produce things without significant outlays. From the knowledge of your fellow hackers to websites such as Instructables, Hackster or Makerzine, hours of videos and discussions are now at your disposal.

This knowledge can then be inexpensively converted into prototypes and proofs-of-concept in hackerspaces, thus allowing people to hop over the initial major innovation challenge: the first product iteration. By taking the risk out of this early phase, hackerspaces are making more ideas viable.

Companies such as FirstBuild will even assist you with the whole process. From the germ of an idea, to improving it with others and the creation of the first prototype, they can even help you to produce your first product in their Microfactory and then sell it on their website. For someone with a good idea but little knowledge or experience in product management, it is a great way to reduce the risk of failing at any of these critical steps.

As I write, the hackerspace community is continually evolving. These community workspaces are integrating into and partnering with incubators and angel investment networks, making it even easier to produce prototypes and on the back of these prototypes get necessary business skills and funding. With this type of support, one is able to launch a product to the world with confidence. By covering the entire process of product development, funding and business skills acquisition, and taking a large portion of the risk out of any new venture, we are seeing more and more innovations growing into successful companies and out of hackerspaces to make space for the next generation of hackers. A true risk lowering innovation.

Chapter 4

Continuously Amplify Your Innovation A Million Fold

Since the time of its discovery, gold has fascinated mankind. Its versatile properties and visual luster have attracted our attention from time immemorial. Although its dominant place in jewelry is firmly established, one may not have expected it to have played a central role in finding new ways to innovative. In 2000, the desire for a modern gold rush became an innovation rush as well. It all began with an unlikely outsider who crushed the mining industry's *status quo*.

Rob McEwen was not your typical miner. Born in Canada, he grew up under a father well versed in the world of finances. His father, Donald McEwen, ran a securities company specializing in investments in the mining sector. Understanding the financial side of the mining industry was part of the family's DNA, and this most likely provided Rob with good intuition for anything to do with managing valuable natural resources.

As a young man, he spent his summers and time after graduating from university working for his father, playing a role in the older man's financial backing of Canadian mining companies. After leaving his father's business for a short stint in the wealth management firm Merrill Lynch, Rob McEwen returned and acquired the family business in the early 1980s. Soon after the acquisition, Rob's father tragically passed away leaving the management and leadership up to the new managing director. However, it wouldn't be long before Rob's new management style ended up uncovering a precious innovation nugget in the company's strategy pan.

In the late 1980s, Rob spotted an interesting takeover target. A mine in Ontario appeared to be a major underperformer. He realized it could be bought for a dime and so he initiated a takeover bid, and in 1989 successfully emerged as the majority owner of the Red Lake mine.

Despite the mine's poor performance, the mining magnate was convinced he'd made the right decision in buying the Red Lake mine. The nearby Campbell Lake mine had already yielded an impressive total of 283 tons of gold (10 million ounces). The puzzling problem was that his mine simply wasn't producing like its neighbor. So far his new pot of gold had only yielded around 100 tons during its operation. Something needed to be done.

Mr. McEwen's belief in the potential of his mine convinced him to pay geologists $10 million to investigate and analyze the mine to find gold seams he was sure existed. A lengthy analysis began which finally provided him with geological reports and mountains of data that could be used to find out where his most promising gold veins were hidden.

It was now 1999, and after almost four years of industrial problems with the mineworkers behind him, the still optimistic CEO had a heap of data and a streamlined operation, but still no certainty about where to launch exploratory digs. It was a fraught situation.

Exploratory digging costs millions of dollars and, if unsuccessful, is about as useful as burying your own money in a large ditch. Hence, a small company like Goldcorp, Rob's company working Red Lake mine, with a market capitalization of around $100 million in 1999, could not afford to take such a risk. Rob realized he needed to try a new tactic.

In 1999, the gold miner attended an information technology seminar for young presidents at the Massachusetts Institute of Technology. It was there he heard a presentation on the impact of Linux's open source revolution on software development. Although the seminar seemed an unusual choice for Rob because the topic had nothing to do with mining, it was a meeting that would write another chapter in the history of innovation.

Linux is a project that people all over the world happily devote their time to creating and improving. It is one of the serious challengers to Windows' dominance of computer operating systems. The key aspect of Linux is the code development carried out by thousands of people—voluntarily.

It was during the Linux presentation at MIT that Rob had an epiphany: why not use this idea and direct people's latent time or expertise to going over Goldcorp's geological data to see if they could discover a buried gold seam? Rob was convinced that by tapping into a global pool of geological talent, he could unearth hidden gems of information in his geological data set.

It was a daring idea. The mining industry is not well-known for being open with its most important information—geological survey data. There is a good reason for this. This data can offer valuable information to competitors and other stakeholders, helping them complement their own data, or providing them the basis for a hostile takeover. Rob McEwen predicted and hoped, however, that by structuring and publicizing a global competition to analyze the data, researchers and geologists worldwide would jump at the opportunity to prove themselves. And he was right.

On March 6, 2000, the Goldcorp challenge was launched. The challenge was open to everyone, even competition-related affiliates of Goldcorp. This meant that anyone, from individual prospectors, geologists, consultants to university geology departments, government mining ministries and agencies, geology associations and so on, could take part. All that was needed was to request a copy of a CD with Goldcorp's full 50-year geological and drill hole database for the Red Lake property. Each entrant could then sift through the data and hypothesize where gold was hidden. The gold rush had begun.

During the four month registration, more than 450,000 people visited the competition website, which was an amazing achievement for a geology competition. To put this stunning number in perspective, at this time only 100 million people in the United States (the country with the highest number of internet users at the time) were even connected to the Internet. Around 1500 teams from 50 countries (as many countries as there are in the so-called "developed world") registered for the challenge, with entrants coming from all walks of life. The response was more than Rob had ever expected.

Mathematicians, geologists, computer scientists, military staff and more, all wanted a piece of the pie. And the pie was large—$575,000 was up for grabs with a grand prize of $100,000 for the winner. But more important than this would be the fame and ensuing recognition bestowed upon the winner and, as many entrants hoped, additional business opportunities.

For the prospective internet fossickers, the schedule was tight. They had four months to submit an initial eight page summary and, if accepted, they then had only six weeks to produce the final report. This breakneck pace was unusual, because such a process would normally take years to complete. The entrants rushed their reports in to vie for a place in the second round, with 25 semi-finalists making it through.

If the first round was fast, the second round was even quicker. The semi-finalists whipped together their final reports within the six week deadline, and presented their detailed exploration plan. The results were to be reviewed by a panel of recognized geological experts.

What was submitted amazed both Goldcorp and the panel. According to the director of the Institute for Global Entrepreneurship and Electronic Commerce, Graham Clayton, the results "exceeded all company expectations."

The final reports fulfilled "not only the standard industry methods, but many new and highly creative applications of advanced mathematics, chemical analyses and applied geophysics."

Despite the pace, the entrants discovered 110 new promising drill locations. The final result? Four out of four of the most promising dig sites struck gold! And the winners? Two small Australian companies that had never even been to Canada.

Since the discovery of these new gold bearing ores, GoldCorp increased the worth of its company tenfold from 1999 to 2004. Its daring idea unleashed a model for innovation in the mining industry, which, despite the significant risk for Goldcorp, succeeded magnificently. The campaign won Goldcorp much fame and recognition from the business community, including being named by *Business Week's* "Web Smart 50" as "one of the 50 most innovative companies on the web worldwide" and being featured in Fast Company's "The Fast 50"—just to mention a few prominent sources.

The entire competition was an exceptional strategic success for the CEO and Chairman, Rob McEwen. It earned him a place in both the gold industry and innovation halls of fame.

What Was It That Rob Did?

Rob's success was due to employing an innovative way to assess his data by using significant participation from 21st century gold prospectors. When the public assists with a problem such as this, it is usually called crowdsourcing.

This term was coined by Jeff Howe and Mark Robinson, editors at Wired Magazine, in 2005. According to them, the definition of crowdsourcing is:

> "The act of a company or institution taking a function once performed by employees and outsourcing it to an undefined (and generally large) network of people in the form of an open call."

We encountered the term previously in the chapter on crowdsourcing, but it is worthwhile looking at some of the relevant concepts again. Crowdsourcing looks to inspire a "crowd" (i.e. the public or some section of the public with specific skills) to take part in a process often with the possibility of being selected as a winner (however it is not always necessary to have a winner). When the process involves a competition, many models of this exist, for example, the crowd chooses a winner, the crowdsourcing agent chooses a winner and so on. In the Red Lake Mine Challenge, it was Goldcorp who chose the winner.

This method of gaining help from a crowd is not new. The ancient example of rounding up "volunteers" to solve the king's wartime recruitment problems was a form of coercing help from a "crowd." The difference now is that the participation is voluntary.

One can view many processes in a more general sense as a form of crowdsourcing. Normally crowdsourcing is a way of using the general public to get services, ideas or content, however depending on what is meant by these terms, crowdsourcing can encompass many things. For example, employing someone can be seen as a form of crowdsourcing with the crowd suggesting possible employees for a role and the employer selecting the most suitable candidate. All crowdsourcing efforts require the person looking to crowdsource something to assess the suggested "solutions" (either on their own or with the help of the crowd) and then choose the best.

What has made crowdsourcing such a recent phenomenon is the Internet. It grants someone looking to fulfill a task the ability (in theory) to contact billions of people easily and quickly. If we look again at the employment process before the rise of the Internet, employers would use newspapers or similar outlets to reach potential employees. Back in 1990, one could possibly reach around a million readers in a major newspaper such as the *New York Times* or *LA Times*. The Internet has made such numbers look insignificant.

The coverage provided by the Internet does not come without its cost. By involving more people with your issue, you'll also have to deal with more people responding to it. It is no trivial task to sort through one million applications you may receive by crowdsourcing something online. Despite this, it was perfect timing for Goldcorp given the increase in Internet usage in the late 1990s, and given the ease with which it enabled them to let all interested parties discover the competition and access it.

For Rob McEwen, the quality problem was not a significant one. He didn't have to sift through millions of answers as the number of people even able to understand the issue was limited. As with any challenge like Goldcorp's, crowdsourcing via the Internet makes it simpler to attract the attention of all the possible experts by placing the problem where everyone can see it.

But of course Mr. McEwen wasn't the first to use crowdsourcing successfully, nor will he be the last. Since 2000, there have been many successful high profile crowdsourcing attempts. Probably one of the best known was IBM's Innovation Jam, which was launched in 2006. It attracted over 150,000 people from 104 countries and 67 companies. It generated 46,000 ideas using IBM's most advanced research technologies. As a result, ten new IBM businesses were launched, with a total of $100 million in seed funding. A number of these ideas have gone on to be successful enterprises in their own right.

IBM took a broad approach to their crowdsourcing competition, but the process can also be quite focused, for example, flying to the moon, taking a photo and then cruising on back home. This challenge was an XPRIZE launched by Google in 2007 to promote the development of private space travel. It featured on the XPRIZE website, which is designed for radical innovations intended to change the world for the better. The prize competition started in 2007 with the announcement of a significant prize for any private organization that could, according to XPRIZE:

- Land a robot safely on the moon;

- Move 500 meters on, above, or below the moon's surface; and

- Send back HDTV mooncasts for everyone to enjoy.

In addition to the big profile crowdsourcing events which gain most of the attention, there are many other projects that have attempted to exploit the efforts of the crowd. Everything from small to large software projects, prize competitions and more have appeared. In fact, around one million tasks are completed by the crowd each day. This is a massive amount of work corresponding to the economic activity of many large multinational companies.

The focus of this chapter, however, is not crowdsourcing *per se*. The process we will look at here is one focused on a company's long term strategy. It is the permanent opening up of an organization's innovation processes to external sources. It is not about having a competition now and then to find expert help for specific problems. It is allowing external contributors to set the innovation agenda, to suggest solutions to problems you didn't even know you had. It is creating an open door to a company's research and development department and finding ways for the company to benefit from this. Appropriately, this strategy is called open innovation.

Open Innovation

It may begin with obtaining help for one problem, then for a second, and before you realize it, strangers are solving all your problems. But companies today don't look at it like this. It is merely a question of attracting anyone with the right skills to offer solutions or suggest new disruptive ideas that could define new markets and opportunities.

Open innovation is a strategy that utilizes external sources of knowledge, expertise and ideas, often in conjunction with internal intellectual property in order to drive innovation in a consistent and structured way. It is the practice of including outside capabilities and resources to expand a company's value proposition in any number of relationships. According to the open innovation experts at 100open.com, it is about "innovating with partners by sharing risk and sharing reward."

There are numerous good reasons to use an open innovation strategy. Many large companies fish in the external ocean of ideas to catch truly disruptive perspectives that they may not be able to catch internally. Companies pursuing an open innovation strategy can leverage their internal research and development to include external ideas smoother and faster than if they had tried to build the expertise in the area.

Often, the first question about open innovation is, just who are the people solving these problems? And the second is, why would I want some random stranger to spam me with his or her new idea for a perpetual motion machine?

In many cases, the people solving your problems are experts. They want to demonstrate their abilities and are looking for a challenge, fame or money. We saw this in the Goldcorp challenge. The participants were geologists, mathematicians and other experts. To have paid each entrant the standard geologist, computer scientist or mathematician consultant fees would have sent the company broke. But they didn't need to. The prize incentive (for many it was fame) was enough to mobilize experts from around the world to participate and do their best to solve the problem.

Although not typical, one form of open innovation could be to run competitions around each innovation focus area a company is interested in. For Goldcorp, and most companies, organizing a campaign to access outside experts is highly advantageous. Not only can they save a small fortune on consultant's fees, but the company running the activity sets the rules. In Goldcorp's case, the competition had a breakneck pace, with only nine months between the competition being announced and the submission of final results. The competition allowed them to obtain an expert solution and to obtain it quickly.

This may sound like a win-win situation. A company gets a difficult problem solved, and people have a chance to make money. However, for the prize winners, the reality is not always so rosy. In the Goldcorp case, the three winning teams each won a total prize of just under $100,000, which was well short of what they normally would have charged had they received a contract to carry out similar work.

In fact, the CEO of Fractal Graphics, one of the winning teams, admitted

> "Our industry has been going through a hard time ... We had been trying to raise venture capital. Any positive news could only be a big help for us."

Despite the fact that the money Fractal Graphics received barely covered the project's costs, they considered it worthwhile and saw the publicity as the main prize. As they said, "It would have taken us years to get the recognition in North America that this project gave us."

If this was in fact their reason for entering the competition, then their strategy paid off well. The Goldcorp competition provided the two winning Australian companies with much valuable publicity and follow-on work for their businesses. In fact, Fractal Graphics had to double its staff and find larger offices to accommodate its resultant growth.

What about the losing companies that also invested their time and money? This was, of course, the risk they took, and is one of the downsides to taking part in such innovation efforts. We will return to this later.

Incentives And Strategy In Open Innovation

So how do you attract the right people to solve your problems and not potentially lose them to competitors? What specific strategies are involved to achieve this?

There are many difficulties facing a company when implementing open innovation. Not least are the two core activities of open innovation—the process of idea generation and of idea selection. There are several questions that arise here: Should one source the ideas from the crowd? Should the crowd select the best ideas? Or both?

When tapping into the ingenuity of the public, one can use a number of strategies. A portal for the reception of ideas is a common one. Companies such as Procter and

Gamble (P&G), GlaxoSmithKline and Shell use this. However, another strategy is to run a competition each time to crowdsource ideas.

Open innovation portals are attractive for a number of reasons. They allow a constant stream of ideas to flow as they arise, companies have tighter control over what is shown when, and how people can participate. Despite the advantages for companies, there are many hurdles for the people generating ideas to get over in order to have a suggestion accepted.

Some of the rules imposed by companies can be intimidating. For example, P&G requires your idea to be already patented before they will use it. Although this may seem tough, it is there to protect the company from being sued for stealing someone else's idea.

In one of General Electric's competitions, they required prize winners to "assign to GE all aspects and content of your Entry, including the intellectual property rights in and to your Entry and all intellectual property." And then further, as though that were not clear enough, they wrote emphatically, "PRIZE WINNERS WILL BE REQUIRED TO EXECUTE AN ASSIGNMENT OF RIGHTS IN ORDER TO CLAIM THE PRIZE."

In addition to being restrictive as to the type of submissions they will accept, and also requiring the legal rights to the crowd's inventions, companies may not jump at the first idea you throw at them. Most companies have a specific list of innovation wants which you may contribute to. So if you have a great idea that isn't what any company is currently looking for, then go start your own company.

For firms with an open innovation portal, there also exists a genuine strategic risk in making their research needs public. To counteract this, many companies state only briefly what the need is. This brevity often extends to many parts of the information provided. For example, there appears to be no company which is prepared to state its profit sharing strategy with inventors at the outset. Typically, most companies consider remuneration for crowd-sourced suggestions on a case-by-case basis.

Given these difficulties, a company may be tempted to go for a public competition. However, it would be unwise to think that because designing a good portal submission system is difficult, the better choice is to use a competition to source ideas. Designing good competitions is also fraught with challenges.

Apart from finding the right incentives, which Gold-corp clearly did, the actual design of the competition is also critical. Everything from the look of the submission portal to simple design decisions—such as whether entrants can see other entries—play an important role. Many websites do not allow entrants to see each other's entries, arguing that if you want to prevent the crowd version of groupthink (i.e. crowdthink) then each submission must be independent of other submissions. In addition, the solutions to the many other issues that arise are also not so obvious. For example, should entrants be allowed to collaborate, or should only direct competition be allowed? It all depends on the nature of the challenge, and you need to think about the pros and cons in each case.

One of the biggest issues is to actually get people involved, and a thought on the mind of any contributor must always be, what incentives are there to prevent the company from just stealing the idea I submitted? That is, why do they even need to pay out the reward once I have given them my idea on a plate? They now have everything they want! To deal with this possibility, innovation platforms have arisen that base their reputation on honesty and fairness, to give people enough trust to submit their valuable ideas. For a beginner wanting to run crowdsourcing competitions, probably the best advice is to use a platform before running one on your own.

Websites like NineSigma or InnoCentive are examples of such platforms, and they allow companies to crowdsource their ideas without having to internalize the costs of open innovation platforms. The websites also allow anonymous challenges allowing companies to keep their identity hidden from competitors. Another benefit is that they can reach more people because they may attract more attention than the company's own website with their varying set of challenges. And of course, tapping into a broader base of participation helps attract a more diverse set of skills. As the latest research shows, this is a critical factor in more powerful innovation.

The next step, once you have chosen the appropriate method to source ideas, is deciding how to filter proposals. Evaluating ideas creates new and additional costs for a company and it may also lead to ill-feeling arising in members of the public (i.e. potential customers) as a result of having an idea rejected. If a company were to do it internally, then the filtering could cost as much as paying someone to come up with the ideas in the first place. To keep costs down, interns or juniors are often asked to filter the ideas, raising the question if great ideas are being overlooked by untrained eyes.

Another way of filtering proposals is to turn to the crowd again and ask them to choose the best ones. This could save a company time and money but once again, it is not without its problems. Businesses such as Threadless, which we will meet later, use the crowd to help gauge T-shirt demand and choose the best designs. However, the crowd is only used as a filter and not as the adjudicator of good design. The final step is completed internally once a set of designs has been shortlisted by the crowd.

Apart from the issues mentioned here, there are many other important details. For example, presenting a problem that is too complex can quickly derail attempts at open innovation. Companies have thus learnt that breaking a complex problem down into more manageable parts produces better results. In addition, providing the right incentives, giving feedback for entries, lowering the barriers to entry and many other aspects need to be fine-tuned before one can expect a successful campaign.

The Ways To Open Innovate

Not surprisingly, there are different ways to engage in open innovation. As the term is usually understood, it means people outside a company contributing to the company's innovation strategy. This is known as inbound open innovation. However, this can easily be turned on its head and a company can contribute to the R&D needs of people outside the company, which is sometimes called outbound open innovation. Usually this involves licensing or selling intellectual property to third parties in the form of patent licensing deals. Although this may sound quite standard, it is far from it.

Companies commonly view their R&D efforts as a competitive advantage. Allowing others access to this, even paid access, would seem like the equivalent of handing over the game plan. This attitude is reflected in the number of companies that engage in inbound open innovation compared to those engaged in outbound open innovation.

In a large study of top European firms, it was found that although around 30% of companies use inbound open innovation, only about 5% use outbound open innovation.

This could be seen as a considerable loss of potential revenue, especially if the new technology is never used. As an example of a response to this, all Procter & Gamble patents come with an expiry date. If they are not used within three years, P&G either "sells, licenses, or donates them to the external market."

Similarly, GlaxoSmithKline has also highlighted and promoted the potential value of outbound innovations "in targeting diseases of the developing world—where there is not the same potential commercial return as in developed countries." GlaxoSmithKline does this "by sharing expertise, resources, intellectual property and know-how with external researchers and the scientific community." Out of this, several projects to fight malaria, TB and other diseases have been brought at very low cost to the developing world.

When it comes to a massive pile of discoveries to sell to others, in 2014, IBM topped the list in U.S. patent acquisition, by successfully taking out a staggering 7,481 patents. And what is the benefit of this expensively guarded pile of intellectual property? Well the United States Patent and Trademark Office's fees are currently about $2000 for searching, submitting and examining an application. If legal fees are added to this, then at a modest cost of $5000 per patent, it means IBM's patent treasure chest set them back almost $40 million in 2014 alone. If this intellectual property is never used, then that treasure chest may as well lie at the bottom of the ocean together with all the wasted money.

In the meantime, other companies, such as Technology Reserve, have recognized this problem and have developed powerful business models to alleviate the patent burden for knowledge-intensive companies. Founded in 2011, Technology Reserve takes a holistic approach to innovation and looks at the end goal of knowledge creation—building products. One aspect of this process are patents and, when leveraged appropriately, enormous value can be gained from them.

Tapping the latent potential of their patents can be a difficult thing for large multinationals. But if a company were to package patents with other useful intellectual property (possibly created elsewhere) combined with an expert able to utilize this package, then the original inventing company or even other foreign businesses may find the raw intellectual property much easier to use in its newly packaged form. Add simple licensing agreements to this and you may have a winning proposition. This is exactly what Technology Reserve does.

Technology Reserve facilitates a way for companies to be both successful outgoing and ingoing open innovators. It provides the possibility to make money from intellectual property, both unused and used. For companies like IBM, it could be the key to unloading their financial patent burden.

Even if IBM were to use these opportunities, the big question for them is, does open innovation pay? To find out, let's have a closer look at a company using this technique.

Procter & Gamble's Story

Procter & Gamble has a long and innovative history. In fact, it started with two men, rather unusually, being invited into a business partnership by their common father-in-law.

William Procter, an English-born candle maker, and James Gamble, an Irish-born soap maker, both immigrated to the United States as young men. Like many adolescents at that time, they were filled with dreams of wealth and freedom. James arrived in the U.S. as a teenager and William arrived 13 years later at the age of 31. Although there was a difference in their emigration dates, both men landed in Cincinnati after their long trip across the Atlantic.

Both continued to practice their trades in America, with James beginning his career in the U.S. as an apprentice soap maker at a local soap and candle factory. William, on the other hand, worked by day in a bank, but still used his candle making skills to earn some extra money on the side.

Then as fate would have it, they both met when they married related sisters, Olivia and Elizabeth Norris. Alexander Norris, their father-in-law, observed that the businesses where his two sons-in-law worked were competing for the same limited raw materials to make candles and soap. He convened a meeting in which he tried to persuade his new sons-in-law to become business partners.

On October 31, 1837, after several years delay, the Procter & Gamble partnership was created. During the 19th century, the firm grew from successful contracts with the military and a committed, diligent workforce motivated by innovative profit sharing schemes. The company kept their pioneering spirit alive throughout the 20th century with the progressive creation of new products such Ivory (1880s), Crisco (1911), Tide (1946), Prell (1947), Pampers (1961), Ariel (1967), Pert Plus (1986), and more. During the period of this innovation stampede, P&G expanded globally, now operating in over 90 percent of the world, producing over 300 brands.

Despite their success, during the late 1990s, Procter & Gamble suffered from periods of lower than expected sales growth. The lack of growth was ascribed to their inability to find game-changing products as they had been doing for the previous 50 years. By the early nineties, it had been over three decades since P&G had seen a breakthrough of the scale of 'Tide' or 'Pampers.' The company realized it needed something new if it wanted to win back its old innovation mojo. Motivated by a desire for growth, the multinational company decided it needed to find a new way to innovate.

Enter Alan George Lafley. In 2000, Alan was elected President and Chief Executive of P&G. When he took over the reins, he decided it was time to change the low tide of innovation. Instead of an ebb of disruptive ideas, it was time to flood the development chain with innovative products. The new CEO proclaimed, "Innovation is everyone's job." To achieve this, a new culture to find and adopt innovations was needed.

So in 2001, Lafley did something unusual for the time. He set a goal of increasing the number of outsourced innovations fivefold, (i.e. he wanted to engage strongly in inbound open innovation). To do so, he threw open P&G's innovation doors via a new organization-wide initiative, called Connect and Develop. Firstly, the initiative defined a set of areas which the company believed to be the most important customer concerns to solve. These foci were then launched both internally and externally via the Connect and Develop platform.

Since its inception, the multinational's open innovation strategy has enabled it to establish more than 2,000 agreements around the world. These efforts have produced a slew of successful new products originating from outside the firm and, according to the innovation experts Rowan Gibson and Peter Skarzynski as stated in their book *Innovation to the Core,* the efforts have simultaneously helped P&G slash its own R&D financial commitment by around 20%. And with a research and development budget of around $2 billion, this translates into a saving of approximately $400 million per year. In 2006, more than a third of P&G's new products and almost half of the ideas in its pipeline were attributed to its Connect and Develop program.

One interesting example from the program is the innovative Bounce Fabric Softener. It was developed and patented by a lone inventor working from home to solve a problem faced by anyone using a clothes dryer. How can one use a liquid fabric softener in a dryer? With a bit of help from Procter and Gamble's R&D department, they came up with a new product to do the impossible: soften fabrics while you dry them in a clothes dryer.

During Alan Lafley's engagement at P&G, he managed to more than double the company's sales, the number of billion dollar brands, and its market capitalization. It made the company one of the five most valuable companies in the U.S. and among the ten most valuable in the world. From an open innovation perspective, it appears to have been a major success.

And The Others?

Procter & Gamble isn't the only company dipping into the open innovation well of ideas. There are many other companies out there fetching their buckets full of innovations. Dozens of large companies are involved in open innovation attempts, many of which have borne fruit for the companies and their creators. For example, GlaxoSmithKline opened its open innovation portal to the public in 2007, which resulted in several significant innovations being developed by the expertise of external partners. The Aquafresh Isoactive product for toothpastes, which according to the company, was "successful beyond original expectations," was developed as a result of these efforts.

Another company that has a long history of being proactive in innovation is Shell. It launched its GameChanger platform in 1996 through which it "seeks out and invests in early-stage ideas that could potentially revolutionize the energy industry." Since its inception, GameChanger has partnered with over 1700 innovators and has invested more than $300 million in over 3000 ideas, turning about 250 of them into reality. However, what has happened to those 2750 other ideas? This brings us to our next point.

What Can Possibly Go Wrong?

As great as open innovation can be, there are ways in which it can all go horribly wrong. This can range from people not liking their intellectual capabilities being doubted, to someone using an open innovation portal to hack the company website. Although these last two examples can be taken care of by good customer relations and decent computer security, other more serious issues are always a consideration.

If you go to GlaxoSmithKline's or Procter & Gamble's website and look at their innovation wish-list, it is reasonably clear what they are interested in. For competitors looking to see what the two multinationals believe the future will be, there's no better place to look than at their open innovation portals. It is therefore clear that companies need to be careful what they reveal when stating an innovation challenge. The correct balance has to be found in providing enough information to entice the right mind but not enough to excite the competition.

Another problem that can occur is the dreaded crowdslap. This is when the crowd parodies the crowdsourcer instead of helping them. For example, in 2007 Chevrolet experimented with crowdsourced advertisements. According to the New York Times, they created a website "allowing visitors to take existing video clips and music, insert their own words and create a customized 30-second commercial for the 2007 Chevy Tahoe." However, things didn't quite go as Chevrolet planned. Instead of developing witty and innovative commercials for the car brand, those who responded assembled 30-second ads that, according to a leading crowdsourcing expert, "Skewer[ed] everything from SUVs to Bush's environmental policy to ... the American automotive industry."

Similarly it is also problematic for companies when people do not take the open innovation challenge seriously for other reasons. This results in poor quality responses that obviously waste time and money to sort through. Whether this arises from a portal being spammed or genuinely bad suggestions, it is a very undesirable outcome.

Finally, open innovation can easily fail if companies mismanage the crowd. This can happen when a company sees an opportunity to exploit the crowd's effort without recognizing it could destroy their users' motivation by showing a complete lack of respect for the goodwill of its participants. In 2014 Yahoo! changed the rules for their photo sharing website Flickr so they could sell photos uploaded by users. Naturally this created quite an uproar and most likely damaged the Yahoo! brand. By doing this and trying to exploit a good relationship, a company can create a lot of bad sentiment among the very people they are trying to encourage to work for them.

How Can I Succeed With Open Innovation?

There are many success stories demonstrating ways the power of the crowd can be harnessed to your advantage. Obviously copying an existing open innovation strategy is not a certain recipe for success, but it is instructive to see, in a general way, which approaches have succeeded.

An informative example is Threadless, which was mentioned earlier. It was founded in 2000 by two entrepreneurs, Jake Nickell and Jacob DeHart, who were both under 21 years old at the time. The idea behind Threadless is that people can submit designs that are then voted on by the Threadless community. It has embraced the open innovation model not only for idea generation but also for idea selection, and this still appears to be working very well.

The first step of their platform allows people to submit a T-shirt design. Members of the Threadless community can then give a submitted design a score between one and five, and those patterns which score the most points are the ones which are printed. To help motivate designers, the T-shirt platform gives rewards for good designs, and pays contributors cash for winning ideas. The business model is a roaring success because community members tell Threadless exactly which shirts to make. Not surprisingly, as a result, every product eventually sells out.

The growth for Threadless has been spectacular. It took only two years for it to have more than 10,000 community members and to smash the $100,000 sales mark. After five years, sales were on their way to hit $5 million. Without any form of advertising, sales force or retail distribution, it grew its revenue by around 500% per year in the beginning, making typical company growth targets of 5% seem meaningless. By 2006, after only six years, sales had reached $18 million, of which profits were $6 million. This would be an unusually high profit margin for any business.

The success of Threadless is based significantly on empowering people to realize their own designs. For Nickell and DeHart it began as a hobby, a side project, until in 2002, it had grown to such a size that it required their full-time attention. From 2002 to 2007, they saw their base of committed users grow tenfold. By June 2007, Threadless had printed 802 designs, submitted by 499 community members, and 68,547 unique submissions had been judged and voted on by the Threadless community. This is a business truly "run" by its community.

These final numbers show that the company's success is based on creating a community where people's contributions are valued, not obviously exploited by the firm. (Although, at the time of writing, this may be changing!) This is similar to other crowd campaigns where people's motivation is increased by recognition, or money, or simply because something is fun. By cashing in on all three, Threadless created a strong and profitable business model based on the almost voluntary work of its members.

This aspect has been critical to the company's open innovation success: the creation of a community around a topic that everyone is enthusiastic about. Forums, discussions and moderated content allow people even more participation and to identify further with the brand. Add to this a company with values aligned with its contributors and you have a perfect match for success.

The complete handing over of the decision-making to the crowd makes sense for Threadless. It's part of the company ethos, and founder, Jake Nickell, says about the community, "We trust them to tell us what is right and we agree with the consensus of the community and adapt to it." Thousands of designs are submitted to Threadless each week and managing them would have required an army of judges. So the company's model is to use its available public to filter out the poor designs and then it only needs a small design team to judge the remaining ones.

The lesson from Threadless is that using open innovation successfully appears to be about engaging people in the right way. This doesn't necessarily mean handing out large sums of money for winners, but it does mean making everyone feel like they are a valuable contributor. This is really no different to well-known principles of good employee engagement, but with these principles operating in the context of an anonymous extended workforce.

After seeing the success of companies like Threadless, many other platforms have appeared that attempt to pursue a similar business model but do it around a different topic. Websites such as iStockphoto, which sells crowd-sourced photos, belong to a list of hundreds of companies trying to perfect the open innovation business model and to provide people with a platform on which to innovate.

If this is something you wish to try yourself, then the main point is to find something that people feel passionate about, or at least very enthusiastic about, and create a brand that listens and is receptive to, and respectful of, the crowd's interest. It is probably not wise to just turn to the crowd to solve one-off corporate disasters which you don't want to solve yourself. Otherwise, as British Petroleum learnt after its attempt to use open innovation to solve its massive oil spill problem, open innovation can quickly become open humiliation.

Threadless, iStockphoto and all the other startups with similar business models offer an illustrative counterbalance to the large corporations we met earlier in this chapter. They show that a startup which embraces open innovation from the beginning can build an extremely successful business. If Threadless had decided to take the significant risk of designing all its T-shirts itself, then it would have entered a highly competitive and low margin market. However, by letting the crowd decide on the best designs as well as creating those designs, it managed to reduce its risk and thrive in a cut throat industry.

The Point

Open innovation has, without doubt, become a common tool in a company's toolbox. If you run a company and are not using it, then the reason it should be in your toolkit is that it addresses the same issue as this book does: the reduction of risk.

Originally, open innovation developed in recognition of and as a way to tackle corporate blind spots. It is just not possible for every company to corner the best minds on every topic. So it makes sense to scale up the brain power of an R&D team by letting people outside the company increase the size and depth of the intellectual pool. Missing those big ideas is a risk and opening the door to people who may have ideas helps to mitigate this.

This new form of innovation has become popular because it provides people with a platform to test ideas and find solutions that are too risky to pursue internally. It also provides companies with a safety check by gauging the reaction of the public to possible new products and services.

A major advantage of open innovation is its ability to attract many different perspectives, which is necessary if a company is to keep ahead on the technology curve. Disruptive innovations are claimed to occur when many divergent perspectives collide, so what better way to facilitate this than by throwing open the doors to innovation? This approach is particularly important for more mature organizations that can develop a risk-averse culture with the focus on efficient operations.

What open innovation is not, is an attempt to cover a lack of ideas with an innovation Band-Aid, such as crowdsourcing a single solution for a single problem. What open innovation is, is a systemic mindset and strategy. When open innovation is incorporated into an organization's innovation agenda, open innovation can be seen as far less expensive and less risky when compared to the cost of the same amount of input from the same number of people had they been employed by say, a company or organization. This was clearly the case for Goldcorp.

However, as we have discovered, care needs to be taken, and the new innovation platform needs to be correctly managed and aligned with community values to pay dividends. Many people may now see the value of those Corporate Ethics courses they begrudgingly sat through during their Business Degrees!

Chapter 5

Choosing A Winner The 21st Century Way

Richard and Mike sit opposite each other in a small room on a university campus. They have just spent the last 15 minutes listening to a rather unkempt man in casual dress explain the rules of a game they will voluntarily play together. It is an unusual game but it seems lucrative: a simple auction where either of them might walk away with ten dollars for nothing!

The game begins. Richard starts and puts in a bid of one cent for the ten dollars. A great deal, pay one cent, get ten dollars. But he's doubtful that he'll walk away with the ten dollars because Mike will surely bid higher. As expected, Mike throws in a bid of two cents.

Richard knows he can't leave it there. According to the rules of the game, both players have to pay their final bid no matter who wins. So if our newly minted gambler stops now, he'll get nothing, but will have to pay one cent. Not bad, but not as good as walking out with a profit of $9.97 if he simply wagers three cents. So he does just that.

However, Mike doesn't hesitate and shoots back with four cents. Annoyed, Richard bets five cents and before he has finished computing his win, Mike has already bet six. Richard ups the ante—fifty cents. That will show Mike that he's serious and hopefully he'll leave the game licking his wounds with a minor loss of six cents.

One dollar. That wasn't part of Richard's plan but now the betting pendulum is oscillating quickly. One dollar-fifty. Two dollars. Three dollars. Five! Richard is sweating. No matter what happens now he needs that ten dollars. A five dollar loss is a day's lunch at the cafeteria, but Mike is aware of that too.

Nine dollars. Richard gasps. If he gives up now he loses five dollars but if he can clinch the win, at least he walks out with something. So he bets nine dollars-fifty, confident that Mike will give up. Before he can relax, Mike bets nine dollars-ninety! Richard scratches his head. There's only a ten cents win left in the game, but if he loses he's out of pocket for two day's lunch. Nine dollars ninety-five. At least he can take home five cents profit.

Then comes the crunch. Mike bets ten dollars. From now on, both know they are losing money. They look nervously at the auctioneer who just grins back. They shoot a glance at each other, knowing who the real winner is. Richard doesn't want to be the biggest loser and decides to cut his losses. Ten dollars and five cents. Surely Mike will give up now.

Eleven dollars. Twelve. Thirteen. Fifteen. Stop! At this point the supervisor stops the auction before it gets out of control. Richard's face is red and he can feel his pulse beating in his neck. His hands are sweaty and he's breathing heavily. This was supposed to be a game. A simple game. An easy ten dollars...

The story above is an example of a famous experiment called the "dollar auction." It is a classic game from game theory, first published by Martin Shubik in 1971. It has been played all over the world with the above results coming up repeatedly. As they progress in the game, the players' behavior invariably becomes more and more irrational, at least according to traditional game theory. Players keep increasing their losses by betting higher and higher.

It may seem like a bit of fun, but sometimes the fun can get out of control. During one executive MBA course, a $20 auction like the one above was held. What were the final winning and losing bids? $100? $200? Try $2,000 and $1,950! The "loser" ended up faring better than the "winner."

According to behavioral economists, what we see here is completely normal. Well, completely expected. Behavioral economists like to focus on how people's decision making deviates from what they would expect from "perfect" decision makers. According to classical economics, a "perfect" decision maker is one who rationally weighs up all the options, decides which gives the optimal outcome, and acts upon it. In considering deviations from "perfect" decision making, one normally looks at those specific behaviors which lead people to deviate from how they "should" behave.

In the dollar auction, players become involved in the so-called escalation of commitments, which is driven (at least in part) by the so-called sunk cost bias. This common bias encourages participants to dig themselves an even deeper hole to get out of than the one they are already in. However, before we delve further into the bias, let's have a look at this pioneering area of economics that tries to explain how human beings really behave.

Behavioral Economics

The field of behavioral economics looks at the real decisions human beings make, in particular the economic ones. Unlike classical economics, which creates models of what the ideal economic agent would do, the focus of behavioral economics is on how less-than-ideal economic agents (i.e. normal people) make decisions in real life. Specifically, how we deviate from a normative model of perfectly rational people. It basically includes all the dirty, non-rational bits, such as emotional, social and other factors, which play an important role in our decision-making. It is these aspects that muddle and spoil the clean mathematical analysis desired by classical economists.

The model of a perfectly rational person has the individual making decisions by sitting down, finding all possible courses of action, comparing the benefits of the different choices and then choosing the one that maximizes either pleasure, wealth or whatever else the person desires most.

The biggest issue with this model for the behavioral economist is not whether the information required for perfect analysis can be acquired, or even whether the eventual choice is right, but the core assumptions about the decision making process. The main problem they have is the hypothesis that people are perfectly rational and able to optimally compute the best possible outcomes for themselves. Behavioral economists ask themselves, do people really carefully analyze all options in detail, or do they use potentially suboptimal shortcuts or other tricks to cut through the whole process in making a decision? And they answer this question with a resounding, "Yes, they do!"

Finding and documenting these deviations from rationality has been a major challenge for economists of the twentieth and twenty-first centuries. Each bias demonstrates a type of behavior that strays from the economic optimum, and usually arises from a shortcut or trick to simplify decision-making. Ever since this new field of economics was started, inserting a long tally of biases after a researcher's name has been a hobby of many academics.

In 1979, while this area was still quite new, a landmark paper was published by Daniel Kahneman and Amos Tversky that laid out a theoretical framework on why decisions deviate from perfect rationality. It wasn't the first paper in the field, but it was definitely one of the most important. In it, the authors established the groundwork for a new model of understanding decision-making, as well as naming and demonstrating some of the first acknowledged biases in behavioral economics. This got the game of bias hunting rolling.

Essentially, the two psychologists outlined a number of common cognitive behavioral biases, which they then used to build a more accurate description of how people make decisions when risk is involved. The two main biases they examined were the certainty effect and the isolation effect. These predispositions were part of an argument that showed traditional utility theory (a theory about how people compare and assess the utility of their actions) was significantly flawed in describing people's actual behavior. One of their more famous results was demonstrating that people are more risk adverse when deciding how to achieve gains than when trying to avoid losses of the same value.

However, a word of warning before you start pointing out to your friends how their behavior deviates from perfect rationality. Although the results of behavioral economics are often presented in such a way that it seems we all suffer from such biases, this is not the case. Even in Tversky and Kahneman's highly influential paper, their results revealed that consistently, between 15 and 30 percent of all respondents don't behave in the ways that have made this branch of science famous, i.e. some people make the perfectly rational choice. So if you can't seem to squeeze your friend into the bias box, then there might be a good reason for it.

The two pioneers' results on biases were made clear, via several experiments, in which people were forced to make risky decisions. Their work proved that people use shortcuts to assist with their decision-making and, according to the authors, these shortcuts are often undesirable and lead to allegedly poor decisions. This opinion, however, is not universally shared. Many high caliber researchers, such as Gerd Gigerenzer, have shown the significant value of certain rules of thumb. In situations ranging from catching baseballs to choosing mates, sometimes our shortcuts can produce much better outcomes than more considered calculations do. This is something we will come back to later.

Despite varying opinions on the benefits of biases, Kahneman and Tversky's work is universally acknowledged as an important contribution to economics. In 2002, they were awarded the Nobel Prize in economics in recognition of their work. This was more than 30 years after their pioneering paper had first appeared and unfortunately it was six years after Amos Tversky's death. Daniel Kahneman, however, made it to Stockholm to receive his well-deserved recognition. He was awarded the prize by the Nobel committee "for having integrated insights from psychological research into economic science, especially concerning human judgment and decision-making under uncertainty."

The importance of this has not gone unnoticed by the rest of the world. Worldwide, countries have started to create behavioral insights teams to help "nudge" people in a desired direction. One of the most famous teams is in the United Kingdom, run by David Halpern. Called, unsurprisingly, the Behavioral Insights Team, they have exploited our knowledge of human biases to encourage political and economic outcomes that more closely match the intent of laws.

Although many politicians were initially skeptical of the value such a team of psychologists and economists bring, their worth is now clear. By simply rewording emails sent to decision makers, the "Nudge Unit" has successfully addressed issues such as minority representation in the police force, by increasing the representation of black and minority ethnic groups by 50%. The team has also improved tax paying compliance by around 30 million pounds, simply by reminding people of their neighbor's tax propriety. Because of their evidence-based approaches, even failures are seen as a success because something is learnt from every attempt to influence behavior.

Based on the successes of David Halpern and his team, Australia, Singapore, Germany, the U.S. and many other countries have now set up similar teams. The UK group's influence has even expanded to the UN, the World Bank and other foreign organizations. Their successful techniques are now being used ambitiously to address even larger issues such as violent crime, unemployment, and homelessness in major U.S. cities.

As behavioral economics continues to achieve significant successes in both the academic and real world, more and more attention is naturally being focused on it. Behavioral techniques and ideas are spreading to other fields, where these evidence-based insights into human behavior are revolutionizing more than just economics.

Behavioral finance was one of the earliest and most natural areas for its application to improve our understanding of money matters. The use of this approach to understand how people make finance and investment decisions has spawned many books, and has also influenced investment funds looking to exploit the way people behave, to either help them make better choices, or simply to make more money. Large funds, touting a behavioral finance approach to investing, became quite popular and successful at the start of the twenty-first century.

But it is not only the finance sector which has bene-fited from the new field of behavioral science. The impli-cations of cognitive limitations and biases in decision-making are as varied as they are vast. The consequences stretch across all aspects of human activity. The one area where this is not just important for scholars, or particular groups, but society as a whole, is innovation.

Decision makers with innovation on their agenda need to comprehend something about those aspects of human perception and of our cognitive processes that can lead us astray. They need to understand something about the psy-chology of choice and its effects on innovation. And it is here that two exciting fields of research have collided and out of them has arisen behavioral innovation.

Behavioral Innovation

In this context, I use the term behavioral innovation to mean the study of the way we make decisions about inno-vation. It relies heavily on the concepts and biases devel-oped by behavioral economics but is now wrapped in an innovation mantel. The new science looks at the way we assess our innovation investments and make choices, whether that be an investment of effort, time or money, in our own or other people's ideas. In this chapter, we will look at ways we can improve the decisions we make about the innovations we invest in.

Behavioral innovation is a new field, still in its infancy. At the time of writing, only a handful of scientific articles deal directly with the influence of biases on the innovation process.

It's important not to confuse this new branch of science with other similar fields. Areas such as cognitive innovation, creativity analysis and the like investigate the actual creativity process itself. The focus here is not on how we create ideas, since I will assume that you have a smorgasbord of ideas to choose from already (if you don't there are many books which can help you with this). What this chapter will help with though, is understanding the biases standing in the way of successful innovation by better recognizing good ideas and how to prudently invest your time and money in them.

For example, one could be considering a portfolio of startup investment options and deciding when to invest in an idea, either as a first time or repeat investor: whether from the perspective of a venture capitalist, an angel investor, or simply your own money or time. Here, you will learn about the general behavioral principles that can unconsciously and adversely, or positively, affect our innovation decision making.

If you have ever invested in someone else's idea, then you will know that making completely detached and shrewd decisions are difficult. It is exactly this challenge the current chapter wishes to address. If you don't think you've ever had this problem, then it is even more important that you read on as you have just clearly demonstrated a potentially detrimental bias!

You may think you can control your behavior and that this doesn't apply to you, but our inherent predispositions lead us to make all sorts of strange and inconsistent decisions. Even decisions you are adamant you will never make and know are morally wrong. In the right circumstances, you won't believe the person in the mirror is you.

In addition, if you think that intelligence or education protects you from suffering from such biases, you are wrong, as these are the very people most often chosen as the guinea pigs of research. It is these future decision makers and leaders who are the subjects of the experiments and who are demonstrating clearly our fallibility. It is typical for people to laugh a little uncomfortably at the decisions other people make in these studies, as if they wouldn't make them themselves—but they often do. When gathering their data on the operation of biases, researchers weren't just using any old college kids—they were some of the brightest university students at the world's top universities.

The following biases are well-established, but their application to the field of innovation is new and again, very useful. They have already been examined in many books in other contexts. I encourage the reader to review these sources and learn more about how to deepen their understanding of the topics touched on here.

One final note: many of these behavioral innovation ideas, like most good ideas, will seem obvious and not need further explanation, but this doesn't change the point that we still exhibit these predispositions despite the fact that we know they can be detrimental to our innovation decision making.

The first class of biases we will consider is so well-known that it requires no introduction and can be pithily termed "the self-delusion biases." However, there are many other well-known and established biases I want to round up under this term. Basically they are all about our instinctual need to create a much better picture of ourselves than the facts warrant. In short, overestimating our abilities and being unduly overconfident.

Are You Deluding Yourself?

Why do so many entrepreneurs enter the high tech startup world even though they are aware that at the end of three years, most of them will fail? In fact, most seems an understatement because the facts point to 92% of them stumbling.

If each person truly believed they would fail, then I'm sure they would invest their time or money elsewhere. Even if someone thinks that having a chance at a financial comet like Facebook is well worth the risk, then the knowledge you will fail many times before getting there (and still, most likely you won't!) and go completely broke along the way, doesn't seem to tip the scales for many entrepreneurs. Hence, just as with driving skills (see below), the vast majority of people, choosing to risk their time and money backing their own ideas, have a completely unwarranted belief in their own ability or chances of success. Or worse, are simply making unreasonable decisions. Claiming that one enjoys going broke, as well as enduring enormous emotional stress, is something I find hard to swallow.

On the investment side, startup investors typically believe that with enough bets, at some point, a few of their gambles must win. The hope is that these winning bets will justify all the losses. Although this strategy may seem to be a reasonable solution to counteract overconfidence in one's ability to pick winners, unfortunately it is not always a good one. It has been shown, that even for seasoned venture capitalists, whose job it is to recognize and select the most lucrative innovative ideas, two-thirds of the projects chosen for funding ended up going broke, or returning less than was invested. Surely this can be improved.

A particularly salient example of a false belief in ability manifests itself in some mutual funds and their generally poor returns. In Gary Belsky and Thomas Gilovich's highly successful book, *Why Smart People Make Big Money Mistakes and How to Correct Them*, the authors compared several mutual funds with simply investing in a stock index. They found that the index outperformed the mutual fund in the majority of cases. Add to this the fact that investing in an index has no fund fees. While some funds can seem to be massively successful, sometimes this is just a sly trick.

To construct a successful hedge fund plan, many banks and fund operators start with a dozen funds, each with a different strategy. These funds are hidden from the public and are nurtured internally in the hope one of them will consistently yield good returns, at which point they are revealed to the masses as the latest and greatest investment available. Just based on simple probabilities, given there are enough such funds, one of them will have to be a winner.

A good analogy is flipping a coin. If you flip it long enough, you're bound to come up with five heads in a row.

In the same vein, eventually one of the funds in stealth mode will have four or five good years and then be abruptly announced to the public as the best thing since complex interest rates on home loans. Little do unsuspecting investors know that it was simply random chance that the fund performed well and not that it had hit on a good strategy. This often becomes clear the following year when the fund plummets and returns a negative yield.

The point is that these funds are run by experienced, savvy and very well paid investment advisors. However, they too are subject to the biases of self-delusion and their results prove it.

Amongst experts, this behavior seems to be quite common. It has been demonstrated again and again that experts are not as coldly rational or sagacious as their high wages would imply. In 1984, psychologist Philip E. Tetlock began a multi-decade study with his current wife and research partner, Barbara Mellers, to investigate the quality of expert judgments of high-stakes, real-world events. His conclusions were particularly revealing.

The data showed that the so-called experts were often no better than chance. In fact, one could predict outcomes much better with simple algorithms which merely extrapolated current and past events. In particular, when the predictions focused on periods three to five years in the future, simple heuristics were the way to go.

The most shocking part of the analysis was the inverse relationship between an expert's prediction accuracy and his/her fame. If that is not a case of self-delusion, then I don't know what is. What's worse, the results were even more endemic in other traditional indicators of foresight. The inverse relationship seemed to hold up not just for fame, but also for self-confidence and depth of knowledge. In Tetlock's book, *Expert Political Judgment: How Good Is It? How Can We Know?,* he presents results from his analyses and classifies his specialists into two classes: hedgehogs and foxes. In relation to the first group he states:

> "Low scorers look like hedgehogs: thinkers who "know one big thing" aggressively extend the explanatory reach of that one big thing into new domains, display bristly impatience with those who "do not get it," and express considerable confidence that they are already pretty proficient forecasters, at least in the long term."

As for the second group, which fared much better, he continues:

> "High scorers look like foxes; thinkers who know many small things (tricks of their trade), are skeptical of grand schemes, see explanation and prediction not as deductive exercises but rather as exercises in flexible "ad hocery" that require stitching together diverse sources of information, and are rather diffident about their own forecasting prowess."

Unfortunately, people are often drawn to hedgehogs, who make bold, overconfident statements and seem to offer us certainty even when it is completely unwarranted.

When it comes to celebrities on TV shows, we are entertained, but when it comes to a brash financial advisor, we go broke.

In summary, the so called self-delusion bias is not something restricted to a small group of people. It seems to pervade society—so beware! The better you think you are, the less accurate your judgment is likely to be. If you are looking at a number of innovative ideas and want to know where best to invest your time or money, then by reducing the effect of this bias, you'll go a long way and have higher chances of making a better decision.

What I have called the self-delusion bias, however, is a combination of other more fundamental biases. To beat this innovation impediment, it is worth looking at each of them in detail so we can see more easily ways to improve our choices.

What Is Going On Here?

The above self-delusional behaviors arise due to many complex, interacting factors, not the least of which is simply a desire to make money—regardless of the means necessary! However, what I am most interested in are the biases that lead people to believe that they can make good innovation decisions despite the evidence showing the contrary.

The first bias I'd like to consider in this context hides behind many names, such as the "Lake Wobegon Effect" or illusory superiority. It is often remembered by the classic poll conducted in Sweden, which resulted in 90% of respondents claiming they were above-average drivers. We can see it in operation in the case of the aforementioned number of failed entrepreneurs. It also raises its under-performing head in situations where every innovation pundit believes he or she can consistently out-perform everyone else. An obviously impossible task!

Another bias very likely playing a role in all of the above examples is an overconfidence bias. The name says it all. People are more confident than is actually justified, which leads them, for example, to invest in businesses they probably don't know enough about (think about decisions being made based merely on a pitch!). People are overconfident that they know enough about the area the innovation is happening in, and can predict the business idea's performance based on that.

Hindsight bias also belongs here. This one is aptly named because it helps prevent people from learning from their mistakes. As the name suggests, this bias is about people either changing their predictions after the fact, or falsely believing they predicted closer to the fact than they did. It basically prevents the requisite feedback loop that enables the development of good "gut intuitions," which are extremely necessary in highly uncertain industries. If you wish to improve your innovation investment record, it is best to learn from your past failures.

Other important biases potentially affecting the above behavior include what could be termed cognitive dissonance. This is a phenomenon that causes us to ignore conflicting information which would contradict our initial, preferred decisions. It allows us to believe two contradictory things simultaneously or simply ignore evidence contrary to our beliefs. It is something that happens a lot in everyday decision-making and gives us the pleasant feeling that we have decided correctly in spite of contradictory facts.

In relation to innovation, this final bias can be devastating; sticking to our guns in spite of evidence to the contrary is a bad strategy. When contradictory evidence appears, if we are not capable of reassessing our previous decisions in light of it, then at least we should ask someone else to do so for us. Apart from the innovation risks, sometimes such poor decision-making can even be illegal, as the directors of many public companies have discovered!

The next and final bias is probably one of the most important. It concerns people who ignore the role good luck and randomness can play in their own activities, such as in the hedge funds we saw earlier. In the relevant literature, this is known as the self-attribution bias. It prevents people from learning and appreciating what truly works. Some claim it is reflected in the roller-coaster performance of mutual funds, which can go from being the best in their class to the worst—all within the span of a year. It also applies to venture capital funds and anyone else trying to pick winners. Believing good performance is solely due to one's superior skills and ignoring random chance is a serious hindrance to becoming better at anything.

Combined with the hindsight and cognitive dissonance biases, we can see that adding the self-attribution bias to our decision-making cocktail makes it a dangerous mix. We completely deny ourselves the ability to improve our innovation decisions by being unable to learn from, or appreciate, the very facts before our eyes. When failures do happen, it is a unique point in time for us to reflect and learn, but the presence of biases obviously makes this easier said than done.

Strategies to beat the biases

It is difficult to overcome these biases on your own, so one suggestion is to obtain many other independent expert opinions on difficult topics. But, at the same time, one needs to be wary of the quality of the "expert's" advice, as mentioned earlier. Having an investment strategy in a venture capital company or any company that relies on one person's ability also relies on that same person's failings. In particular, when the stakes escalate and it becomes a matter of quality and not quantity, then beating these biases is critical.

To overcome biases, we need data. In particular, the self-attribution bias can be quickly dislodged by keeping a close eye on the facts. It is important to sort out what part of your performance is due to randomness and what part to skill, easily accomplished by comparing each decision you make with one chosen randomly. If you're not beating the coin toss, then that tells you something.

In general, it is a good idea to spread your bets. In fact, this is a winning strategy against most biases. For example, don't develop or live in an investment system where you bet everything on one investment. By having strict investment rules for startups or other projects, you can reduce the impact of overinflated confidence in any single idea.

Finally, making sure the necessary feedback loops are operating is critical. Over the long term, it will improve your decision making abilities and so lead to more accurate decisions, which means less risk and more money left over for further investments.

In summary, data, numerous independent opinions and risk spread over multiple investments are powerful ways to lower your innovation risk. They may sound familiar because they are well known strategies for many purposes outside of behavioral innovation. The problem, like everywhere, is simply sticking to them.

That Sinking Feeling

During the 1980s, IBM was the poster child of the American corporate world. It had just launched the PC, which was revolutionizing the home computer experience. The computing giant had also closed a fantastic looking deal with a little known startup called Microsoft, to provide it with an operating system. Together they created a PC with the MS-DOS operating system, which became an icon for the era.

MS-DOS was a command line based interface, which meant no pretty windows or mouse controls. Users had to type what they wanted the computer to do, in plain text on a black screen. Despite their commercial partnership being the envy of other computer producers such as Apple, IBM saw more long term potential in developing their own operating system. They thought they would take off and leave the scraps to the rabble at Microsoft.

Although financially successful, the hardware-software pair weren't winning their customers based on functionality and ease of use. This accolade soon went to Apple with its fancy graphical interface controlled by a mouse. The Microsoft/IBM duo, however, conquered the market on price. IBM had created a highly modular computer architecture for which Microsoft could build a cheap operating system.

Despite its commercial success, by the mid to late 80s, both parties in the partnership saw the need for a new visual interface. In the ensuing years, Microsoft had been busy quietly developing another operating system. At the same time, IBM had Microsoft help them with an operating system by enlisting their expertise to build and release the OS/2 operating system in 1987.

The OS/2 software was a significant improvement over earlier IBM operating system versions, however, it lacked what people wanted most: a Windows-style graphical user interface. Unfortunately for IBM, at the end of 1987 Microsoft released Windows 2.0—with much better graphics. This became the turning point for IBM because it marked the beginning of a decline into what would become a financial decision-making fiasco for the PC maker.

As people started moving to Windows *en masse,* IBM scrambled to get out an operating system with a Windows (or more accurately, Apple) style appearance. A year later in October 1988, IBM fulfilled expectations and released OS/2 version 1.1. Everything was looking promising for Big

Blue, until it was discovered you couldn't print with it. Instead of saving themselves, they put another, of what would be many, expensive nails in the OS/2 coffin.

In a quick succession of releases, Windows continued to produce new operating systems, which were simply eating up IBM's market share. At the same time, startups such as Dell and Compaq began beating the International Business Machines Corporation on its own hardware turf. The real innovation in IBM's PC had been the open modular design. Computer chip makers such as Intel had capitalized on this and built a range of processors that could be easily used to put together a computer. This in turn opened up the market to other low cost competitors for IBM, such as Dell and Compaq.

Determined to stay in the game however, IBM marched bravely on and launched more and more versions of OS/2. In 1992, in spite of clear warning signs of impending disaster, the PC pioneer launched OS/2 version 2.0. Coincidently, just as the company started to hemorrhage cash, it was well into a partnership with Apple to develop yet another operating system called Taligent. In retrospect, this was a strange decision, but for IBM it was to be the operating system to end all operating systems. Unfortunately, it didn't end that way.

In 1992, IBM choked and coughed up a nearly $5 billion loss. But their ailment had just begun. In 1993, they topped their 1992 loss with another $8 billion withering away. Big Blue's CEO John Akers was swiftly given the axe as they belatedly realized that continuing to spend so much money to keep a dead business model alive was only killing the company faster.

What happened to this pillar of the American corporate world? IBM management had fallen for one of the greatest bias traps around; the sunk costs fallacy. As they committed themselves further to one bad operating system after another, they blinded themselves to the objective reality that their product was just not working. They didn't want to pull the plug on the losing product they had already invested so much in. Instead, they let more money sink down the drain as a result of their bad decisions and spent far too long pursuing what they refused to see as a lost cause.

The sunk cost fallacy is driven by previous investments, whether they were good or bad. It causes us to justify future investments on the basis of our past ones. This can be bad news if it leads us to deepen our cash or time contributions to a startup or idea only because of earlier commitments and not because of merit. This can be especially dangerous if we are reinvesting to prop up our struggling fledgling. You don't want to be the only one sitting in a deck chair on a sinking Titanic trying to get the most value out of your ticket.

Another bias called the endowment effect also contributes to our inability to let go of a bad idea. This bias makes us value or appreciate more what we have done or put effort into than an equivalent amount of work done by someone else. For example, we want more money for our own hard work than we would be willing to pay if someone else offered the same service to us. When combined with the sunk cost fallacy, it means we will keep on investing in what we already have an investment or contribution in, and this effect is strengthened by the fact that whatever investment there already is, we will subjectively overvalue it. The end effect makes us think that what we already have is worth more than the value given to it by the rest of the market. Unfortunately, the market is actually the determiner because it is the one paying for the final product via a merger, an acquisition or a floating on the stock exchange.

The endowment effect can stop us from realizing appropriately and at the right time that what we are really dealing with has little value. It makes us overvalue something and so hang onto it longer than what may seem reasonable, throwing more gold into the coal furnace. Typical examples of this can be seen in many startups. We invest our time, effort and money, and then hold onto them sometimes a lot longer than is reasonable. Whereas an outsider, who doesn't have the same bias, would say "I give up," many an entrepreneur continues in an untenable situation only to fall harder than they would have if he or she had given up earlier.

Another major factor contributing to the habit of throwing good money after bad is the loss aversion bias as discussed earlier; people try to avoid the same value losses more than they will try to make the same value gains. This means they are more willing to take larger risks to avoid a loss than to achieve an identical gain. So as people lose more and the stakes increase, they are willing to bet more to gamble themselves out of a loss, à la IBM.

It seems that when we are in a bad situation our intuition and biases try to help us out by giving us the proverbial posterior boot. Big help! These biases, which probably helped save our ancestors in situations of life or death, now end up killing our innovation judgments. It's not that these biases are completely useless, just that when faced with certain decisions, if we end up with a worthless innovation investment on our hands, our biases will make it harder to drop the lump of coal.

The critical thing to recognize is that the successful fruition of an innovation does require investment. Whether it be time or money, there are always decisions to be made as to how much and when. So it is important to be on top of the sunk cost fallacy, endowment effect and loss aversion when looking at reinvesting, whether in startups, new ideas, or simply your own business.

Strategies to beat the biases

In the venture capital industry, companies and startups need to reach specific milestones before being given more money. If they don't, then they are mercilessly weaned off their funding teat. Having strict investment criteria like this can prevent you from even allowing yourself the possibility of reinvesting in bad eggs, because you already have fixed rules in place to stop it.

As easy as it may sound, this has its downside. Such a method prevents over-investing in lost causes, but what if the product or service you are developing is only one step away from a breakthrough. Is it not better to put in the last few cents to see it over the line? The uncertainty caused by this can make the previous strategy difficult to implement.

Not only can it be difficult sometimes to know when to quit, but perseverance can have collateral benefits. For example, in the late 1980s, Apple decided to develop a personal digital assistant. It would be able to take notes, manage contacts and a calendar, plus much more, together with dedicated handwriting recognition. Although Apple pushed the product for well over a decade, the Apple Newton was a miserable failure.

In spite of the device being completely unsuccessful on a commercial level, it taught Apple an enormous amount about small, mobile, hand-held devices. In fact, some claim that if Apple had given up on the Newton idea altogether, then they may not have had the basic experience to create the massively successful iPod, iPhone, iPad and other related products. So it's not always certain that pulling the plug on a bad investment is a good thing. Maybe lessons can be learned from the burnt finance?

Although I claim that entrepreneurs need a reality check on when to give up on failing ideas, how many of our greatest companies would have died in their early years if they had folded to the bleak outlook at the time? Often entrepreneurs say they only succeeded because of their tenacity. So this advice is not as clear cut as one would like but it doesn't mean the situation is hopeless.

The fundamental strategy to beat these biases is to try to limit potential damage. This can be approached from the venture capital perspective above, but you can also achieve this by limiting the possible amounts you can lose via things such as diversification. So even if one innovation experiment fails, you have others to prop you up. In addition, an investment in one idea is then always compared to others. So if you only have a fixed budget, then the opportunity cost of each investment decision will be weighed up. So if you necessarily cut a company off just before its critical breakthrough, then you can be happy knowing that the other opportunity you chose was better anyway. The final advantage is that spreading your bets also limits the amount you can invest in any one idea and if more money is necessary, it will force you to find others to share your investments. This may diminish your gains but also your losses.

Another way of dealing with the previous biases, like the sunk cost fallacy or endowment effect, is to forget what we've invested in. That is, every time we want to invest more in something (an asset), we need to study the investment like it was the first time we ever heard about the project. This allows us to assess innovation targets on their merits, irrespective of whether we have invested in them or not. A practical way of achieving this could be to ask someone else to do a blind review of the case without knowing the current investments in it. This can also be done to avoid the endowment effect in relation to things we own (think of a problematic car). Again a trusted, competent, independent opinion can be very valuable.

Strategies to beat the sunk cost fallacy and endowment effect can be summed up in a single sentence: save yourself money by saving yourself money! And there is no better way to lower the risk of your innovation investments than by limiting your financial losses.

The Animal Instinct

As the lion gazes hungrily at the herd of gazelle from behind a bush, it is sure of one thing: if it can separate just one of them from the pack, then lunch is guaranteed. As it bolts across the narrowing gap between the gazelles and its anticipated meal, the herd springs to life and gallops away. It's now a strategic game of maneuver and counter-maneuver. The first move is to confuse and scatter the group, so that a sick or young gazelle becomes isolated and therefore an easy target. It's a strategic game of pursuit, only in this case the stakes are life and death.

It is not just African animals which flock together for protection or support. Forming herds seems to be a common characteristic of many life forms on our planet. It must have appeared early in the evolutionary tree because of the large diversity of creatures which display this behavior. Insects, animals and, most importantly, humans, all demonstrate this basic instinct. Protection and power come in numbers and we, as well as many other life forms, intuitively understand this. In fact, according to the renowned Russian scientist Peter Kropotkin the most successful species are those that provide each other with mutual aid and support.

Our desire to form groups has probably helped to keep human beings on their evolutionary branch more than once. Languages and cultures around the world reflect this, and many races have their own version of the "Safety in numbers" saying. In Japanese, it's *Kazu okereba anzen nari:* "If there are great numbers, there is safety" and in

German, it is *Gemeinsam ist man stark:* "Together one is strong."

However, when making decisions in the modern world, herd behavior doesn't always seem to be an advantage. In fact, it has become recognized as another decision-making obstacle, and joins the long list of biases, unsurprisingly under the names of herding or conformity bias. Basically it means we have a strong tendency to blindly follow the group without applying enough of our own independent thought or analysis, and this behavior might lead to suboptimal results.

We see this all the time in financial markets when stock bubbles burst, with people losing their retirement savings. Bubbles are characterized by unjustified exuberance or confidence in an asset (share stock, house, lump of gold, etc.) and when this happens on the stock market, it is the group mentality that gives rise to the bubble in the first place. Some even claim that the great financial crisis in 2008 was an example of a herding crisis. Before the bubble burst, many independent voices were pointing out that the situation was untenable, even those heavily invested knew it, but the crash happened anyway. We all held hands and jumped over the financial cliff together.

This final bias is of critical importance in relation to innovation, because we need to both use it and be able to ignore it at the relevant times. Recognizing the herd bias can help us think independently when we should, but also help us to follow the group when it makes sense! As much as everyone likes to think that they are leading the technology trends, the fact is most of the time, we will be following one. This is a critical skill for both the individual and for a company, i.e. the ability to identify important movements and not miss that trend train.

An example of this recently happened to two of the world's biggest communication firms. As the smartphone market exploded between 2007 and 2010, Nokia and Blackberry were caught flat footed and paid dearly for it. Initially, the herd instinct for both these companies was to keep with the large existing market of mobile phones and not follow the breakaway class of smartphones. Neither of them took the trend seriously enough, nor did they react quickly enough. Although other biases and poor foresight may have also been playing a role here, Nokia went from controlling 40% of the global market in 2008, to under 5% five years later, and by then they had become a takeover target (and were subsequently bought by Microsoft).

Understanding the relevance of the conformity bias or a group mentality and its resulting behavior doesn't just allow us to recognize real innovation trends and follow them promptly, it is also important for our crowdsourcing efforts, for brainstorming and much more. As stated in the New York Times bestselling book, *Wisdom of the Crowds*, the power of the crowd comes from many independent points of view. So if you want to innovate with modern techniques of idea collection, then it is important to work against herding behavior.

In summary, although herding might have saved us as a species in the past, when it comes to innovation, it can mislead us. There's nothing wrong with following the group when it's appropriate... the hard part is figuring out when this is the case.

Strategies to beat the biases

One of the better strategies to deal with this bias, is the same one that helped with the previous two: try and get an independent second opinion by someone who is not involved in the matter. Someone who knows nothing about a trend's popularity or what interests are at stake can provide a much less biased opinion as to the merits of the issue.

While it would not be a good strategy to always ignore emerging popular opinion or popular demand, companies such as Nokia and Blackberry moved too slowly on the smartphone ascension. They stayed with their existing products and ended up either losing everything or almost everything. This is why a second opinion is recommended, to avoid either staying safely with the group when change is better or imminent, and to avoid blindly pursuing a direction no one will ever be interested in. An independent second opinion can provide a safeguard against both these pitfalls.

Changing our focus slightly, there has also been a significant amount of research undertaken to help reduce the effect of a group mentality undermining situations like brainstorming sessions and other related activities. Breaking the bonds of group think is important for companies where innovation investment decisions are made by a group within the company. A technique developed by Robert Epstein, from Harvard University, called shifting, has been developed to break through this mental gridlock.

This approach involves brainstorming as a group for a period of time, and then taking a break to allow people to consider the ideas on their own before returning to the group again. This simple trick has been shown to significantly improve the group's creativity by giving people a break from the peer and social pressures of group dynamics. Adding a trick like this to the decision makers' investment analysis can obviously improve the quality of the chosen innovation targets and at least gives an opportunity to break free of the herd mentality.

The Point

As we have seen, biases can have a profound effect on our innovation decisions without us even being aware of it. When looking at opportunities to pursue particular ideas and technological changes, we need to realize there is a lot going on unconsciously in the background of our minds helping to form, and at times misinform, our decisions. By understanding this and being aware of these blind spots, we can make better decisions and reduce our innovation risk. If fewer of our innovation attempts fail because we are making better decisions, then we can be more confident when investing, and more confident that what we pick will more likely be winners.

In addition to the biases presented here, there are other cognitive issues playing a role in separating good decisions from bad ones. By no means is the above list of biases meant to be an exhaustive list of obstructions decision makers face. Another major cause of poor decision making is often old-fashioned stress. For example, in Dan Ariely's critically acclaimed book, *The Upside of Irrationality*, he demonstrates, not surprisingly, that under stress we tend to make worse decisions. This means important decision-making should never be rushed and should be allowed enough time for proper consideration, and preferably not done when we are under pressure.

We have also only covered a handful of the many known biases. Those discussed here were chosen because they are well-known and relatively easy to explain. We could continue by analyzing the over 100 other biases documented to date. However, this would become tedious and the general concept would be lost in the detail. I have focused on the particular biases which can lead people or companies to the most expensive and least gratifying result of innovation, namely bankruptcy. The challenge and the aim here is to have a good working knowledge of the most important biases and use them to improve our innovation decision-making.

Another important thing to remember when looking at biases is that they are not always bad. It is easy to see biases in a negative light when they are taken out of their natural habitat and placed in the world of, say, high risk finance. Those same biases which would have saved you from being eaten by a lion are now costing you the lion's share of your finances.

Continuing on this theme, it is worth noting recent research, which proposes that the best decision tools we have are our intuitions and heuristic thinking (hence biases!). For example, Gerd Gigerenzer has created a schema to identify when our heuristics can be useful. He freely admits that if the risks are known, and you are well endowed with facts, then jumping on the bias train will probably take you to the wrong station. Here logic and statistical thinking reign king. However, where the risks are unknown, or uncertainty reigns, then good decisions also require good intuition and smart rules of thumb. As he writes, "Calculated intelligence may do the job for known risks, but in the face of uncertainty, intuition is indispensable."

This doesn't mean, however, that someone who has never had experience with startups should intuitively start advising billion dollar venture capital funds. Intuition needs to be developed and refined, and in many situations this is not even possible. Feedback loops with some form of predictable cause and effect, to garner experience, are critical to developing an effective gut feeling.

As an example, take the so-called recognition heuristic. This operates by making us choose the name we recognize when faced with a question we have little knowledge about. One famous study compared German knowledge of American geography with American knowledge of the same. When asked which is bigger, Detroit or Milwaukee, only 60 percent of Americans correctly answered that Detroit is larger than Milwaukee, whereas around 90 percent of Germans got it right. The same result was demonstrated when the roles were reversed.

So one can sense an obvious tension between the different approaches in certain areas. This tension is driven by the different perspectives on heuristics, i.e. for certain behavioral economists biases drive rational failings, however, for certain cognitive psychologists, heuristics can be safer and more accurate than a considered calculation. The trick is to know when our biases can be our best friends and when they are our worst enemies, and as stated this depends on the level of uncertainty in a situation.

The discussion above is there to help you recognize your limits. As explained in the fantastic book by Gerd Gigerenzer, *Risky Savvy: How to make good decisions*, heuristics, or shortcuts, can be beneficial if used appropriately—although they require one to know their limits and not suffer from the previously mentioned self-delusion fallacy. To reiterate, if you have no experience with something, then applying heuristics (your gut feeling) where there is no sensible feedback to learn from, is a recipe for disaster. Alternatively, computing optimal strategies in situations where the risks are truly unknown, will only leave you calculating your own overconfidence.

For those of you wanting to decide which camp suits you best, it would be remiss of me to let you believe that researchers are cleanly split between the two possible perspectives. Gary Klein, another famous psychologist who demonstrated some pioneering results with firefighters in the 1980s, also believes in the power of our intuition. However, in 2009 he co-authored a paper with Daniel Kahneman where they elucidated their different perspectives and agreed on "the issues that matter." So there is no need to wear one hat or the other. Both approaches present valid perspectives, each useful in different situations.

Improving our ability to make unbiased decisions will clearly reduce the level of risk involved in our innovations. It will stop us from overinvesting, provide us with a more logical valuation of our investments and grant us perspective when we need it most. If, however, we still find ourselves succumbing to intuition, at least we can console ourselves that when stranded in a life or death situation on the savannahs of Africa, this skill would probably serve us better than a behavioral innovation optimized one!

Chapter 6
Free Knowledge For Free Innovation

Every innovator's dream is free access to pertinent knowledge. Knowledge is the catalyst speeding up innovation, endlessly recombined and reused in novel applications. If we want cheaper and faster paced innovation, we should increase the availability of knowledge and make accessing it as easy as possible. Unfortunately, the way our world works contravenes this in many ways.

Not only have university fees around the world skyrocketed (though exceptions like Germany do exist) but other ways to access knowledge have become unaffordable as well. Every year, billions of dollars in taxpayers' money is being put into research around the entire globe. Once this research has been completed, it needs to be shared with others to have any value. Under the current publishing paradigm, however, this involves submitting the results to journals for publication, which can then demand thousands of dollars for access to published articles. The upshot? The average person cannot afford these knowledge-based innovation catalysts even though the research was paid for with their taxes.

From the publisher's perspective, this arguably outdated system works well. And the best part of this model for those publishing the research is that, with the recent advent of digital publication, distribution outlays have sunk to practically zero.

Since the initial research is funded by someone else, this part costs the publisher nothing. Also the necessary peer reviewing of the paper is carried out by other scientists and is usually free for the publisher. Nowadays articles can be put online after publishing, also costing the publisher very little. Some journals offer editorial services, but not all. So why is the price for an article so ridiculously high for so many research periodicals? Perhaps it reflects a world view when education and literacy was the domain of the wealthy minority. Perhaps it is just profit mongering.

For example, at the time of writing, publications such as the Journal of Comparative Neurology have annual subscription fees of around $30,000 for both the print and online publication. These preposterous fees have mobilized significant opposition as well as the rise of new publications that seek to make the peer reviewed research more accessible.

According to many people, this lack of access to knowledge holds back innovation and progress. It has provoked opposition from librarians, scientists and even politicians. Their stance provides inspiration to hundreds of thousands of people around the whole world to join forces and pry open the gates to humanity's intellectual achievements.

The emancipating power of knowledge is not a new concept. Mahatma Gandhi advocated the importance of intellectual freedom in *Hind Swaraj* published in Gujarati in 1909 and supported it with the book's copyright notice, which read: "No Rights Reserved." But the connections to Gandhi run deeper. Just like the famous Indian's tumultuous life, open access has had a turbulent journey. Unfortunately, to get where we are, one of the 21st century's pioneering technologists had to tragically die like the leader of the Indian Independence movement. But before we look at this tragedy let us start with the basics.

Open Access

The business of scholarly publishing is highly lucrative. In 2013, it was valued at ten billion dollars, which is about a third more than what the social network behemoth Facebook earned in that same year. This is serious big business for an industry that caters to a small part of society (the highly educated), and it is an industry vigorously defended by publishers to protect their profits.

In fact, a vicious battle is currently raging, involving media giants clashing with vociferous librarians. On the frontline is an unlikely man acting as one of open access's unofficial leaders. What is both even more impressive, and ironic, is that while he is currently vocally fighting for our right to easier and more reasonable access to knowledge, years earlier he almost lost his voice to Grave's Disease. This man is Peter Suber, born in Evanston, Illinois in 1951.

Unorthodox and critical, Peter seemed to develop social awareness at an early age, leading him to become an open access advocate. His father was in the jazz publishing business and his mother an Illinois state senator. One can imagine the effect her powerful character and influence had on Peter given that she was the first woman slated by the American Democratic Party for state-wide office.

The formative values in Peter's background pushed him towards an eclectic set of apparently opposing careers. After first beginning to work as a comedian, Peter dramatically altered his course and accepted an offer from his alma mater, Earlham College, to become a philosopher. This would prove to be a life changing decision for him as it was through his work as an academic that he realized the full potential of the Internet.

The new philosopher had an epiphany in the early 1990s after publishing an academic paper on the Internet. Despite the Internet's small user base at the time, he quickly received emails from other academics wanting to discuss his ideas. Peter decided to focus his attention on the new communication medium as he recognized the revolutionary potential it offered the publishing world by being able to connect and collaborate with minds anywhere in the world.

Around the same time, one of the most prominent research-based open access forums appeared. Initially dedicated to physics, arXiv was a free scientific online archive started by the physicist Paul Ginsparg in 1991. It began as a pre-print repository for physics research—a place where the draft version of scientific papers can be placed before submission to a peer-reviewed scientific journal. The trend of uploading pre-prints of articles has spread to many other disciplines and nowadays, arXiv includes papers from disciplines ranging from biology, finance, mathematics and computer science.

ArXiv and the accessible nature of the Internet appealed to liberal academics. Activities such as these were reaching the point where they were becoming more mainstream and seen as a way of being able to break down the old hierarchical knowledge structures controlled by powerful, overpriced journals.

And these efforts seem to be succeeding.

In 2000 it was estimated that around 20,000 articles were published and made freely available. Nine years later, that figure grew to around 200,000. Even if the number of new journals is taken into account, this increase is still significant. The momentum has gathered pace in all areas of publishing with the venerable arXiv already archiving its one millionth article in 2014.

Given the interest for open access already in the academic system, Peter joined the movement at an ideal time. Peter had been a part of the open access movement for almost a decade, when in December 2001 he played a decisive role in the movement as an advocate at an historic open access conference in Budapest.

The importance of the conference in Hungary's capital should not be understated. From this event arose the Open Access Initiative, which officially defined what open access is. In the mandate, open access is defined as:

> "Free availability on the public internet, permitting any users to read, download, copy, distribute ... the full texts of these

articles ... or use them for any other lawful purpose, without financial, legal, or technical barriers."

In Hungary, the historical document outlining the definition of open access was drafted. Peter's vision and intellectual prowess helped guide the creation of this crowning achievement and under the original 13 signatories, Peter Suber's signature stands proudly. By the end of 2015 this document would have attracted the endorsement of more than 5,900 individuals and over 800 organizations.

Open access embodies the idea that peer-reviewed, academic research should be accessible to anyone with a connection to the Internet. The original premise was to make journal publications more accessible, but it has now expanded and is being applied to monographs, theses and scholarly book chapters.

The definition seems simple; just give everyone unfettered access to research results. The implementation means that articles are published and visible for all. However, already at the meeting in Budapest there was a noticeable rift in views as to how this should be achieved.

By 2001, there had developed two fairly well established, but competing, proposals on how to provide open access: green and gold. These two ways of providing access split the supporters into two camps.

On the one side, many authors argue that the only sensible way of achieving freedom of knowledge is to make the articles freely accessible in a repository outside the publishing journal. This is called green open access. The other camp believes that the author submitting an article to a journal should pay a fee (although sometimes nothing) to assist in subsidizing the costs of publishing. This would allow the research periodical to provide free access to whoever required it. This is called gold open access.

Although these are not the only means of providing open access, these two methods have dominated the discussion ever since 2001. Apart from the work to refine the principles, there have also been many national memorandums of support for open access. In 2013, numerous mandates and other initiatives were started in the U.S., UK, Canada and other countries to try and force more open access to publicly funded research.

As a result of these developments and Peter's determined efforts, the U.S. Department of Energy (DOE) released a document on its "Public Access Plan" in July 2014 with a platform to "offer distributed full-text access to all DOE-affiliated accepted manuscripts or articles after an administrative interval of 12 months." The National Science Foundation has also joined the DOE with a new mandate, according to its website, to have research results "deposited in a public access compliant repository designated by NSF." This program began in January 2016.

Research bodies in the UK and Canada have followed suit by mandating new rules for authors receiving money from certain research funds. They are required to make peer reviewed journal articles available in the author's institutional repository or any other easily accessible location.

The Canadian Institutes of Health Research was actually the first North American public research funder to mandate open access, having adopted the practice in 2006. Their other two funding bodies, the National Science and Engineering Council and the Social Science and Humanities Research Council joined the Institutes of Health Research to create the Tri-Agency Open Access Policy. They released their final policy publication in February 2015 with similar access rules as the U.S. Department of Energy.

On hearing the arguments for open access, one might reflect on what the concrete benefits of open access may be? These include the possibility for faster and broader assimilation of relevant cutting edge research; the acceleration of scientific breakthroughs made possible through easier knowledge sharing; providing access to a public good; and as has been demonstrated, open access publications increase the number of citations a work receives—something all researches desire almost above all else.

Given that the benefits of open access publishing are clearly substantial, does any of this result in what Peter Suber would have hoped for? The resounding answer would seem to be yes, and one of the celebrated success stories in open access publishing has been centered on advances in the fight against cancer.

The Cancer Game

When Jack Andraka left his mark on cancer research, his peers had just started noticing the opposite sex. At an age when most people are being driven by hormones, 13 year old Jack was more concerned about the fact that 40,000 Americans were dying from a particular form of cancer. His was an admirable concern, allegedly aroused by the death of a close family friend.

In 2011, the young pupil was attending North County High School when he became interested in pancreatic cancer. Whilst most university libraries are struggling to keep their journal subscriptions, schools for prepubescent boys and girls are lucky to have a decent selection of novels, let alone medical journals. So Jack was forced to turn to open access literature and sources such as Wikipedia and Google to research his new found passion.

Jack's goal was to create an over-the-counter test for pancreatic cancer. To do this, he combined an insight he'd had in his high school biology class with cutting edge carbon nanotube technology. He tenaciously wrote to around 200 labs before being given access to the equipment he needed to test his hypothesis. With access to a lab he could now realize his dream of creating a cheap and simple cancer test.

It was a struggle for a 13 year old—weeks of recondite internet research, hundreds of rejections, isolation from his peers—but to Jack it was worth it. As his research claims, the technique Jack developed is allegedly over 100 times faster, tens of thousands dollars cheaper, and is hundreds of times more sensitive than other leading tests.

Jack's story, however, is not the only example of how open access research has led to advances in the fight against cancer. Given that this disease is one of the greatest killers in western society, responsible for nearly one in three deaths, it is natural that people outside the official medical research community would try and find answers to stop cancer's insidious spread. Sometimes even a naïve approach and a background in computer security can lead to impressive results.

Matthew Scholz is a software engineer by training. Before reorienting his career compass, he had enjoyed a successful livelihood as a software entrepreneur. However, in 2008 he decided that it was time to change course and began his own independent research into immunology and gene therapy. He had no formal training in biology, no degree in medicine. He just had a desire to solve the riddle of one of humankind's most debilitating diseases. To his aid, he had access to free journal articles.

To make significant progress in cancer research, it seems one often needs innovative and unorthodox thinking. Scholz's previous work experience gave him such an advantage, and he quite naturally utilized computer science tricks in his approach to the cancer problem. His rationale being: if computers can adapt to cope with a new software virus, why can't we do this with human cells?

So Matthew launched headlong into his newfound passion. After encountering one hurdle after another, he decided on a different approach to solving his than simply going it alone. Instead of researching and solving each new problem himself, he consulted experts to do it for him. "Every time I ran into a question I couldn't answer, I'd recruit a new advisor," he stated.

However, it wasn't just the access and utilization of experts that made it possible for him to start a company set up to solve the supposedly impossible: it was made possible by open access journals and Wikipedia. This kept his limited funds available for attending conferences, allowing him to acquire even more advanced knowledge in order to tackle the second leading cause of death in America. Despite the Immusoft founder receiving help from consulting experts, soaking up the latest research at conferences and partnering with several receptive institutions, without basic access to information he could not have even begun his inspiring work.

At the time of writing, Immusoft has raised $2.7 million and had successfully completed tests on mice. The new CEO is now hoping to use the newly raised funds to carry out human trials. If they are successful, he will seek approval from the U.S. Food and Drug Administration to then commercialize his treatments. As Alan Leong, an analyst at Biotech Stock Research states, "The odds are against him probably getting this far". However, without free access to the necessary knowledge, he could not even have rolled the dice.

Immusoft may not succeed in its stated endeavors, but the fact that it is even possible for non-experts in biology—like Jack Andraka or Matthew Scholz—to educate themselves enough to start a venture or potentially generate a breakthrough treatment for cancer can only be seen as testaments to the value of open access and its ability to lower the risk and barriers to participate in solving humanity's problems.

Another important avenue of access to our accumulated knowledge, particularly for those without the means to buy publications or easily access the world of online journals, has always been the library—a long standing, traditional institution. However, the ability of libraries to offer their invaluable service as a knowledge custodian is becoming increasingly threatened. Subscription costs are bankrupting them.

The Serials Crisis

Libraries are the vestibules of our knowledge. For millennia, they have preserved and guarded the written records of mankind's achievements—sometimes even at the cost of the lives of those guarding these hallways of wisdom.

One of the greatest and most influential libraries of the ancient world, the Library of Alexandria, was destroyed after an edict by Emperor Theodosius I in CE 391 made paganism illegal. This library is said to have contained many important and famous works of antiquity, including the Homeric texts, mathematic and scientific scrolls. This knowledge turned to ash in the incineration and was lost in the ruins of the temples.

When it comes to destroying the foundations of modern day knowledge repositories, there are two words that cause most librarian's spectacles to steam up: serials crisis. Although ransacking and torching buildings are mostly a thing of the past, libraries of the 21st century are still under attack, but this time the attack is focused on their budgets. In many developed countries, instead of library budgets being increased—as would seem the natural response to widespread education—many budgets have remained stagnant or even been cut. Simultaneously, the prices of books have often risen faster than inflation—journals being some of the most egregious price setters.

Double digit journal price hikes are quite common in an industry where, on average, the prices increase by almost ten percent annually. With such price increases outpacing inflation by more than a factor of two in most developed countries, this business model amounts to little more than a commercial scam.

If a journal or any product becomes too expensive, the natural market reaction is to stop buying it and thereby boycott such price rises. The publishers of overpriced journals have long recognized this and have developed a strategy for overcoming it. It is called bundling.

Libraries do not need to acquire every journal on offer. Typically it is the few high quality ones that matter. And so how can a publisher sell the lower quality publications? By bundling their flagship journal together in a subscription with other overpriced, sub-standard journals.

One could argue that all knowledge is valuable, but for journals this argument is not true. To better understand how bundling works, we need to know what makes a journal's content more or less valuable.

To determine the value of a journal, one simple measure is the number of citations it receives. If, for example, you publish a paper and the writer of a later published paper quotes your paper, then your paper has received a citation. When someone cites a paper it is an indication that the research in that paper makes a significant contribution to the foundation upon which further research is built. It is a simple proxy for the importance of scientific research.

So if a paper has thousands of citations, it means thousands of people think the article is important. If an article receives no citations, this most likely indicates that the research was irrelevant to other researchers. Citations are useful because they help acknowledge and distinguish well regarded and useful research—hence the landmark distinction of peer reviewed journals.

Many publishers have seen the potential for a lucrative business in publishing research articles. Their business model is to ruthlessly extract as much as possible from the system and create numerous journals that publish large quantities of valueless and unscholarly material. This substandard product is then peddled to universities and libraries for large fees. Libraries and universities are able to subvert this scheme and avoid paying for such worthless products by looking at the citation count of a journal in order to avoid purchasing the poorly cited ones.

In order to calculate the value of a journal, libraries with a fixed budget look at the price per citation they receive for subscription to a particular journal. The lower the price per citation, the better is the value of the journal. A high price per citation means a journal is likely to be publishing substandard research just to make money, or simply charging disproportionately high fees for the good stuff.

Unfortunately, comparing and determining a benchmark for good prices can be difficult. Not only do publishers use the aforementioned bundling trick to bind poor and high value journals together but they have also developed a bunch of legal tricks to pry open library coffers.

When selling journal bundles, publishers often force their buyers into secrecy agreements concerning the prices they pay. Luckily, a group of bold researchers recently made details of these contracts public by invoking America's Freedom of Information Act and obtaining copies of them from several large American Universities. What came out was revealing.

To determine whether a publisher was overcharging the university, the researchers established a benchmark price per citation and compared the data to this. When all was said and done, it became clear just how much certain publishing houses were overcharging for their journals. The price per citation for many journals was roughly ten times the benchmark! So instead of paying an affordable $100,000 for a bundle of journals, libraries are being overcharged a million dollars.

Obviously, with such large sums of money involved, those with vested interests vigorously opposed the aforementioned memorandums for open access. Between 2003 and 2008, the Reed Elsevier Group alone spent over $23 million to lobby the American government! And it was just one of many publishers lobbying governments on open access issues.

Libraries and passionate individuals around the world continue to fight for open access. Their organizations are getting larger and their actions better coordinated. We have seen major successes in the last few years because of ever increasing dissemination of information on the Internet. The currently available 10,000 journals now providing open access to the latest research are a testament to its feasibility.

The serials crisis is caused by something antithetical to open access. In fact, the serials crisis is not only bankrupting our libraries but our innovative entrepreneurs as well. Where certain companies' short term profit goals cause money to be unnecessarily spent on knowledge acquisition, this money is clearly then not available to be better spent elsewhere.

Open access also creates a far wider network of people able to contribute to difficult problems. By allowing articles to be disseminated for free, one is creating a community of knowledge sharing, which is invaluable. This has already provided a strong precedent for the benefit that can be gained when researchers are able to work together and share new knowledge. Indeed, it was a community of open access minded individuals who solved one of physics' greatest challenges of the 21st century: what holds the universe together.

The Greatest Open Access Story Ever

Physics is in a crisis. After the accomplishments of 20th century—dubbed "the century of physics"—physicists have found themselves in the embarrassing situation of realizing something is amiss with the so-called standard model.

The standard model is one of science's greatest achievements. It describes the fundamental building blocks comprising our universe and the interactions between them. By the end of the 20th century, scientists recognized that this scientific edifice was missing something. And this something was very fundamental: the mystic force holding our universe together.

Dubbed "dark matter," this missing matter that prevented the universe from expanding was puzzling physicists and causing a dilemma at the heart of the science. But it wasn't the only problem they had.

Two of the most important pillars of 20th century physics, quantum theory and the theory of relativity, seemed to be irreconcilable. After over a century of digging deeper and deeper into the nucleus of physics to find answers to these mysteries, it was decided to delve even deeper.

For a significant part of the last hundred years, scientists had been smashing atoms and other elementary particles of matter together in order to discover more and more about the universe. The problem was that to understand what made the previous discovery tick, you needed to build an even bigger machine to slam things into each other with even more force.

Hence, one particle accelerator after the other was built to achieve this goal. The most recent contender in this ever mounting challenge is the biggest device to smash atoms yet, the Large Hadron Collider built by CERN, the European Organization for Nuclear Research.

As far as organizations go, CERN is impressive. It manages the largest particle physics laboratory in the world, is the World Wide Web's place of birth, and has five Nobel prizes associated with it. To top it off, their Large Hadron Collider contains the world's largest refrigerator and, according to LHC Machine Outreach, is able to "hold 150 000 ... sausages at a temperature colder than deep outer space."

CERN's mission is to promote fundamental research, and a major part of that has been firing elementary particles at each other at velocities close to the speed of light. They have been building particle accelerators for over 50 years in their attempt to understand more about the universe. Their latest is their biggest and has the most ambitious goal so far.

Everything about the Large Hadron Collider is big. It alone receives $1.3 billion per year, which is more than many countries' entire research budget. To achieve its goal of finding the elusive particle holding the universe together, it spent around $14 billion to build the facility. It has become the world's sixth most expensive research project and in terms of costs is only behind projects such as the International Space Station and the Apollo Program.

The Large Hadron Collider consists of a massive circular tunnel 27 km long crossing the Swiss-French border four times. It has been dug into the ground, with the deepest part almost 200m below the earth's surface. Construction of the particle accelerator was approved in 1995 but as the first test ran 13 years later, there were immediate explosions but not of the expected type. During the inaugural tests, a helium leak led to a disastrous explosion damaging dozens of specialized magnets. The repair work and accident analysis set the whole project back more than a year.

It was in November 2009, when the world's biggest particle accelerator started shooting protons and electrons through its long corridors that the data deluge began. The amount of data created each year was on the order of tens of petabytes, or about 200,000 DVDs of data.

This amount of data is far too much for one person to analyze. So, in order to draw conclusions within a reasonable timeframe, the information is sent out to be analyzed by thousands of researchers worldwide. This data is then used to make significant discoveries, which are duly published to the world with a corresponding amount of fanfare. One such result was the discovery of a long sought particle that was hoped to be the final missing piece in physics' standard model.

After three years of tests and analysis by researchers around the world, the preliminary discovery of the fundamental element, the Higgs Boson, was announced. Its existence allows scientists to now explore the so-called Higgs field. This field is believed to create the basic force holding everything together. The official press announcement at the CERN research center in Switzerland was a momentous occasion, with the man who predicted the Higgs Boson, Peter Higgs, being present to witness it. The pronouncement received extended applause, as well as tears, from many in the crowd.

The work and flood of data did not stop there. More analysis was carried out over the ensuing year to confirm with even greater certainty that the discovered particle was indeed the long sought after Higgs Boson. In a typical cinematic display, Hollywood even featured this elusive particle, before its official discovery, in the 2009 film Angels and Demons, starring Oscar winning actor Tom Hanks.

As important as the collaborative work on analyzing the data is, being able to access the results by all parties working in this field is critical to the project's success. And as quoted by the designated Chair for the Publication Committee, Professor Tony Doyle,

> "When you are working with 3,000 collaborators from across the world, open access certainly does make life an awful lot easier. In fact, I don't think we could do the work we do without it."

All scientific papers produced that analyze elementary particle data are published open access with the data also being made available by the Open Data Portal. This type of access to experimental data and the conclusions drawn from it are what rally the world's best physicists to this unique project and allows faster discovery and verification.

At any point in time, there can be as many as 60 articles circulating amongst researchers and being discussed for their implications. The only way such high powered science can work is when all parties have access to vast masses of information and the inferences drawn from it. In fact, the open access mindset creates an enormous and engaged community driven by the principles of sharing and inquisition. It has been an integral part of the Large

Hadron Collider's and its European Union funders' operating principles ever since one of CERN's employees created the World Wide Web to make better communication and collaboration possible.

However, the fast dissemination of scientific results is not just necessary for scientists working with the Large Hadron Collider project. To guarantee the future supply and quality of good physicists, one must be able to expose students to the latest research. By showing the discipline as an exciting, growing subject, one has a better chance of recruiting more potential physicists into the halls of this venerable science. The success of this has largely been secured by the free access for lecturers and teachers around the world to these ground-breaking achievements.

CERN's use of the open access model has been a major driving force of innovation in physics. The ability to disseminate and analyze new results as they become available is seen as one of the pillars of CERN's success. Its progress can be viewed as one of humanity's major achievements, with open access powering the day-to-day work. Allowing anyone interested in taking part in these discoveries to access the necessary information for free is a major innovation catalyst. It lowers the risk to practically nothing, so that anyone with something to contribute to the discussion can easily do so.

Unfortunately, not all open access stories finish on such a high note. Open access has also had its share of tragedies amongst its highly charged and committed supporters.

Open Access's Tragic Death

An activist, entrepreneur, talented computer scientist and much more, at the youthful age of 26, Aaron Swartz had already achieved much and had a promising future ahead of him. However, this halted abruptly when he was found dead in his apartment on 11 January 2013. He had tragically hung himself. At a point when the open access movement was still celebrating its momentum and successes, it took the necessary time to pause, and consider what had led to the death of one of its greatest advocates.

Aaron Swartz showed a precocious and promising aptitude for computers from a young age. When he was only 13 years old, he won the prestigious ArsDigita Prize for creating The Info Network. This website was an unintentional clone of Wikipedia but clearly demonstrated the young boy's foresight and innovative thinking style. Then just a year later, he co-created the RSS web feed format which is commonly used to generate continuous news headlines and is now used by millions of websites. Either of these would have been an impressive achievement for an experienced coder, let alone a teenager.

Apart from his internet related achievements, Aaron was also a strong political activist. He promoted progressive goals, helping launch the Progressive Change Campaign Committee, demanded political accountability via websites such as Watchdog.net, and also fought for the open access cause. He was also instrumental in blocking the proposed *Stop Online Piracy Act.*

The proposed *Piracy Act* was a bill introduced in the United States with the goal of granting United States law enforcement agencies the power to tackle, among others, online copyright breaches. The part of the bill that made many people nervous was its power to require internet service providers to block access to infringing websites. If this were to be misused, no website would be safe from exclusion, and therefore freedom of speech could become subject to spurious political control.

The bill seriously concerned many people including the young internet activist who campaigned to have the new laws blocked. The movement garnered the support of over 115,000 websites, some of which, such as Wikipedia, went so far as to shut down their services. Their efforts were successful and by early 2012 the bill was sidelined.

Aaron spent his time fighting for many causes such as this, using whatever methods he deemed appropriate. However, after a string of high profile mass downloads, in late 2010, the prodigious programmer sealed his own fate. Over the course of a few weeks, Swartz used a research fellow position at Harvard University to download around 4.8 million research articles. He did this by way of a Massachusetts Institute of Technology JSTOR journal archive account.

Probably unaware that he was under suspicion, Mr. Swartz was arrested in early January 2011 by MIT police and a United States Secret Service agent. He was charged with two counts of breaking and entering with intent to commit a felony. However, it didn't stop there. Six months later, he was charged with wire and computer fraud, unlawfully obtaining information from a protected computer, and recklessly damaging a protected computer. A year later, federal prosecutors added nine more felony counts to his credit.

Aaron was facing up to 50 years in jail and fines of up to $1 million. Whether due to the stress, or the threat of being incarcerated for such a period, Aaron tragically committed suicide on 11 January 2013. In a sad twist of fate, the charges were dropped after his death.

This young prodigy, like Peter Suber, had been a key figure in the open access cause. His vocal support for open access and the activities he organized drew helpful attention to the movement. The "Guerrilla Open Access Manifesto" is probably one of his best known writings, a passionate plea urging people to "fight back" against the "private theft of public culture."

By downloading and publishing around 2.7 million federal court documents online, he was also responsible for putting the Library of Congress's complete bibliographic dataset online for anyone wanting access. This meant the average citizen could circumvent the expensive services offered by the Administrative Office of the United States Courts to American citizens wanting copies of court documents. Despite the fact that they are not peer reviewed science, these documents are still sold in a similar way as research journals. At the time of writing, each page costs eight cents even though the documents are not even copyrighted and contain matters of public record, which are theoretically freely distributable.

Aaron contributed in many ways to a greater public understanding of, and commitment to an open web, and hence open access. His passion for open access added significantly to its cause and increased the ability of people around the world to access the necessary knowledge to advance our common knowledge-based innovation agenda. His death was an untimely one and had this not happened one can only surmise what he may have achieved. The Los Angeles Times wrote:

> "He used his prodigious skills as a programmer and technologist not to enrich himself but to make the Internet and the world a fairer, better place."

The Point

Fast innovation requires easy access to knowledge. Open access addresses this basic need. It is one of the many steps involved in successfully realizing an idea as cost effectively as possible, and hence is an integral part of low risk innovation. The interplay between crowdsourcing and open access is also interesting and very close; for open access to thrive, one needs many scientists willing to publish their results to the public for free.

A movement that seeks to increase freedom of access to knowledge can help support an innovation revolution by empowering the development and dissemination of knowledge, which is often a risky and expensive part of any undertaking. More results are becoming accessible to a larger swath of people. Open access is reducing the cost of access to knowledge, and in doing so, making that money available for other purposes. It is returning knowledge to the people paying for it (taxpayers in many cases) and permitting amateur scientists and entrepreneurs to do things previously thought impossible.

As in any human endeavor, there are problems to overcome or to be alerted to. When it comes to the quality of some of the work published, the open access movement has been criticized. In 2013, the world's premier journal *Science* carried out a test by creating machine-generated versions of a fake research paper and submitting it 304 times to different open access journals. The paper was accepted by a shocking 157 journals.

The result is even more problematic when one acknowledges the author's assessment of the article:

> "[a]ny reviewer with more than a high-school knowledge of chemistry and the ability to understand a basic data plot should have spotted the paper's short-comings immediately."

Although this seems to be somewhat damning for open access, for many people the result demonstrates instead the failure of the scientific peer-review process.

The question is to what extent the result was caused by a journal's access policy? On the other hand, it is well known that certain traditional publishers, especially ones that run conferences in desirable locations, have also published highly dubious "science" articles. This clearly shows that readers of any type of journal need to use discrimination, but at least with free access, readers do not have to reach for their wallets as well.

Another criticism leveled against open access is directed principally at the gold access model, whereby authors pay journals to publish their articles. Apart from creating a conflict of interest in that publishers might publish anything for a fee, the gold access model could make it much more difficult for under-privileged academics to publish. By forcing them to pay to publish, scientists and researchers in developing countries or in underfunded departments, typically the arts and humanities, may be unable to pay the fees and hence be forced to cut research output. As stated by Achilleas Kostoulas, a top language researcher at the University of Graz in Austria, "[Gold] Open Access will likely make it easier to consume research outputs, but harder to produce them."

There are many publishers keen on extinguishing the open access fire before it properly blazes up. However, their arguments and protests are often based on pseudo-science and fears, and not facts. For example, the fear that open access will lower subscriptions, is simply not justified. There is ample evidence that indicates the claim is not true. The American physical Society and Institute of Physics Publishing, two highly respected societies involved in the publication of scientific results could "not identify any losses of subscriptions for this reason and that they do not view arXiv as a threat to their business ... rather the opposite."

To balance the more controversial issues surrounding open access, the success stories continue to appear. We have seen how it helped to drive new cancer research, new fundamental understanding of our universe and much more, so we can only hope that the trend continues. If we want to empower the billions of people without access to expensive books, journals and well-funded libraries to be part of a groundbreaking innovation effort to solving the world's most pressing problems, then we need to give them access to the requisite knowledge. Open access provides a way to do this.

Chapter 7
The Innovation Services Revolution

There is a cloud on the horizon. For the last 10 years it has been looming larger and larger, engulfing more and more of the digital sky in front of us. The gathering forces are part of an industry pivot which is not just changing, but completely revolutionizing the way we interact with our digital devices, and simultaneously making the virtual lives of consumers easier and more reliable. Not only that, but it is making innovation cheaper and more powerful.

The revolution is driven by cloud computing and its unique ability to offer software products by way of a service. Pay for what you use; not for what you might use, as has been the industry norm with software products in the past. Cloud computing allows one the ability to break down the costs of something large, complex or expensive into smaller, more affordable amounts spread out over time.

The advantages of cloud computing are not just cost based. The defining features of the new services are almost ubiquitous access, on-demand self-service and rapid elasticity to respond quickly and efficiently to consumers. For those developing tools built on a cloud platform and sold via a cloud-based service provider, such as Apple's App

Store, the distribution and reach of such services are second to none.

Since the initial development of this form of interconnected computation, many new incarnations and spinoffs have appeared. Today, almost everything is being offered via the cloud, even the proverbial kitchen sink. Not only have existing products such as Microsoft Office been extended to the pay-as-you-go model, new infrastructure is accessible via the cloud—for software platforms, manufacturing hardware, desktops and much more. It is all part of a trend to reduce the barriers of entry into the market for many industries, both new and old.

Cloud services makes these products and services cheaper by reducing the upfront costs and hence increasing the number of consumers and level of engagement. At the same time, if the price of a company's latest offering can be reduced, its business pitch can be made more compelling to investors and customers, leading to more successful business launches. New services aside, a cheaper product means more potential customers, which means more potential income. Hence, business models based on cloud computing have given birth to thousands of startups; companies which would not have existed otherwise.

All these new cloud services have spawned a vernacular of terms and abbreviations. For example, it is quite common to see four letter abbreviations starting with a capitalized letter and followed by aaS, e.g. IaaS, PaaS, SaaS, FaaS etc. The general abbreviation theme is XaaS, where X denotes a cloud service, for example Software, and the aaS stands for as a Service. It is this cloud based business model which is creating the next software revolution.

Both enterprise and ordinary consumer focused companies are driving a wave of growth in this area, with around $2 billion of venture capital pouring into cloud computing in the United States each year since 2011. In 2008, the top 10 public cloud companies had a total market capitalization of less than $25 billion. In 2015, that had risen to more than $170 billion. Almost 700% growth in just seven years!

Who could have predicted such staggering growth? To have done so, of course, would have been highly lucrative. If you'd been able to identify this trend from the beginning, your bank balance would now have risen to somewhere white and fluffy.

The Beginnings

The basic principle behind the idea of offering something as a service, instead of an outright purchase, is not new. For example, renting an expensive object rather than purchasing it is common practice. Rental agreements have allowed people since antiquity to put a roof over their heads without paying the full cost of the building. By letting out a place of residence to someone for a moderate fee, you lower the barriers of access to a home for many.

Another prominent example is the recent trends created by companies such as Uber and Lyft which have turned the focus to no longer owning a car but just requesting it when needed via an app. Viewed this way (and including taxis which have been around a lot longer), one can claim this as a form of Car as a Service.

Seen in this light, cloud-based business models are not new. Despite claims of being a new computing paradigm, the present system of offering computer packages as a service, instead of purchasing them outright, is almost as old as computing itself.

When people first started building computers, they were large clunky devices like mainframe computers. To compute something simple, it often involved piles of punch cards or other memory-preserving devices to store results. Even inputting data into these older technologies followed a similar laborious process. Not only were these early machines slower than current mobile phones and much more difficult to use (it's common to see photos of toddlers taking selfies nowadays!), they also required a whole team of experts to maintain and operate the system if you were going to derive any value from them.

The original room-engulfing machines were the privilege of well-endowed governments and private institutions. This meant that only the very wealthy could afford to buy them. Recognizing this as restricting their potential and a far larger customer base, computer mainframe companies began to offer less expensive options of hosting business applications on these electronic giants. This was known as time-sharing or utility computing, and the initial business model grew by allowing organizations a place to store their data.

The concept of hosting other people's programs seemed such a powerful one that as early as 1966, John McCarthy, one of the founders of artificial intelligence, even thought that time-sharing computing "may someday be organized as a public utility." We are not at this point yet, but as most cloud-based service providers offer free versions of their tools, we have, to a certain extent, achieved a similar result.

Following utility computing, the next big splash in the service pond was the arrival of the Internet—the force driving the current groundbreaking business model innovations. During the 1990s, a number of companies decided that a lucrative way to offer services via the Internet was to host third-party software solutions online. Access to these services was provided to a user by requiring them to download certain programs onto their computer.

The programs located on a user's computer were called the client side software. This software was used to connect to the Internet and leverage the services provided by the respective company. The companies offering services via client side software and an Internet connection were called Application Service Providers.

As the Internet continued its inexorable march forward, with improvements in infrastructure, web-browsers, HTML standards and much more, more sophisticated business models became possible. Websites appeared which offered storage for your photos, videos and other multimedia, rich blog experiences or simply a platform to monitor and facilitate your communication with friends.

These developments continued to one logical conclusion: the Software as a Service business model.

Much of this paradigm shift was possible simply because the user could take advantage of the services by using nothing but their web browser; no more client side installations, updates or backend divisions for different customers. Everything was seamlessly rolled together into a single web-based experience not requiring any more effort from the user than typing in a web address and logging in. The operating system, computer type, etc. no longer mattered.

This final step lowered the barriers to usage and opened up new possibilities to clients who considered themselves amateurs in the world of computing. Many interfaces became so easy to use that practically anyone who knows how to operate a computer could use the service offerings. Whole new markets were thrown open, just by simplifying access to existing services.

With each new wave of change, companies adapted by offering their services according to the new computing paradigm. What led to the hyperbole and opening of checkbooks in the early "noughties" was the mushrooming of billion dollar companies sprouting up everywhere offering these services. Accordingly, when Drew Houston and Arash Ferdowsi, the founders and CEO and CTO of one of these new billion dollar startups called Dropbox, launched a company riding the hot new trend, it was ironically a return to one of information technology's earliest business models!

Dropbox is a company which has tried to change the way we store our files online. Its concept began with an extremely simple and intuitive way to upload and synchronize files amongst multiple users or devices. Combined with the ubiquitous freemium model, allowing users to test many features of the software with limited storage space, they have successfully grown their business to over 100 million users. This growth has also led to widespread success for the company due to their clever and intuitive products. The goal is to seduce with a fluid interface and give more and more incentives to store all sorts of files in the cloud until the point is reached where you have no space left and you need to pay.

When Dropbox appeared on the scene, the idea of storing files on someone else's computer was not new. There were already several existing cloud storage service providers, such as Box, Amazon Web Services and Microsoft, which allowed people to store almost anything in the cloud. And even before that there were many services granting people the ability to store specific things, such as photos and videos online. So when Dropbox entered the starting blocks in 2007 it was joining an already highly competitive race.

Drew and Arash's reasons for starting Dropbox, however, were novel. Many file storage services used unnatural and complicated ways to manage files. Restoring older versions of a deleted or changed file, or simply updating a file, seemed unnecessarily complex. So the founders asked themselves, why should cloud storage be difficult? We all know and understand file storage systems like those on our PCs or Macs, so why can't our Internet backups be similar?

Dropbox was launched as a straightforward answer to these simple questions. It was nurtured in the famous startup camp, Y Combinator, to help it grow the business idea. Despite missing their university graduation, the founders graduated from the camp with some capital in their pockets and officially launched the storage service globally at the 2008 TechCrunch50 annual technology conference. They went from success to success, but a large part of what they did was only possible because of the Software as a Service business model. Crucially, this model allowed them to cheaply start the business and grow with demand instead of launching one large, expensive product which would have been very risky and found a far less positive reception.

Dropbox's subsequent growth also highlighted a gap in the market for such services. This is now being filled by numerous other companies offering storage as a service in varying formats. Due to the file storage data rush, many companies have entered a race to the bottom with the leaders in the field continuing to compete for users and market share. In spite of this, it is still an attractive market to enter. In 2016, the Software as a Service sector will be worth more than $100 billion, or roughly the size of the global toy market.

The "Cloud"brian Explosion

The last 10 years have seen a rapid increase in both the types of services and the number of companies playing the cloud game. With the stunning rise of the Internet arose the possibility to easily offer online services to a mass market. Between 2008 and 2015, the entire "Anything as a Service" industry grew more than 20 times over. Contrast this with the fact that during the same period many industries experienced little or no growth due to the financial crisis in the USA and the rest of the world. As a product, these solutions have become so popular that it now seems, as of writing, that almost every large software company offers cloud-based services.

This growth is not only driven by software companies writing millions of lines of code, but also by voracious market consumption. A 2014 study of over 1,500 information technology and security decision-makers revealed that the majority of the respondents' organizations had already begun migrating to the cloud. The primary reason? Increased agility. According to the respondents, cloud-based services can give companies a compelling competitive advantage.

As Internet connections become faster and faster, more and more users are being enticed into the realm of cloud computing. Services which once required significant bandwidth to be feasible are now readily accessible, paving the way for disruptive businesses based on faster data transmission.

Previously, the main cloud-based offerings were for things like simple storage and viewing, so long load times were acceptable. However, this has completely changed and customers now expect interactive websites and viewing without lag. Entire industry sectors such as online movie streaming, which were once impossible, are now common. In addition, other data heavy applications, such as visually rich design projects, can now be carried out in the cloud via services like Canva. Designers can use Canva or other graphics heavy programs via the Internet and still enjoy a similar experience to doing the work on their own PC or Mac.

With the increased ease of use available by simpler access to ever larger bandwidths, many new businesses have taken an old concept and "cloudified" it. Everything from accounting to client relationship, management to marketing and even healthcare solutions, now have their own cloud-based versions. Instead of managing work on a single indispensable computer, employees can access their data from anywhere on any device, creating the unenviable situation where one is always available and has no excuse for not being able to access their work.

At the same time, the introduction of cloud tools into the workplace has caused a whole set of new issues to surface. Security has become a critical matter and is an essential part of a company's cloud computing suite, making the entry into this new software world a complicated one for large enterprises with significant confidential information. Successfully switching a business's IT activities now requires a bird's-eye view as it is necessary to launch all parts of the solution simultaneously. For large companies, something which was supposed to give a competitive advantage can suddenly transform into an albatross if they are unable to get the transition right on the first try.

The other major challenge for companies is to figure out where to even enter the cloud panorama. One now has a plethora of choices, ranging from simply developing programs and putting them on a platform, to leasing the hardware to build your own platform and customizing it to your needs.

The Expansion Of Cloud Services

As the cloud computing industry has grown, so too has the variety of customer choices. When an industry matures as the online services one has, there naturally arise numerous sub-specialties within this now broad concept. Currently there is no clear consensus on the myriad quantity of cloud sub-categories, but several market segments stand out from the rest.

These newer market segments have now become a battlefield for the giants of information technology, fighting tooth and nail for market share. Older companies, such as Microsoft and IBM, which should have stood to gain the most from the new service provisions, were caught flat-footed as the clouds rolled in. At the time of writing, they are fighting their way back into what should have been their markets after having been disrupted by startups and out-of-industry players, such as Amazon.

Amazon is a great example of a company driven by its founder's, Jeff Bezos, relentless desire to take advantage of the Internet. What began as a company selling books online has developed into a global goliath which, towards the end of 2015, superseded Walmart as the most valuable retailer in the United States.

As Amazon aggressively expanded its service offerings into products other than just books, at some point, it began selling people unused time on its computer servers. In 2006, the company launched its Amazon Web Services and began creating a portfolio of web based products.

This new set of Internet products seemed to have caught Amazon's competitors off-guard, or perhaps it simply showed that the old players were not as ambitious as the quickly expanding bookseller was. In 2015, it was estimated that Amazon's cloud revenue had eclipsed the combined cloud revenues of Google, IBM, Microsoft and Salesforce.com.

Over time, Amazon has expanded its web services and now offers customers many levels of access to its computing resources. These services have been developed under the umbrella of Amazon Web Services and the products have evolved into an important part of the company's business. In 2015, they contributed significantly to Amazon's profit, already making up one sixth of the Internet booksellers' revenue, and rising.

Amazon has now penetrated almost all areas of the cloud market. If we start at the lowest service level and work our way up, using Amazon's services as a guide, we can get a good overview of all the different cloud strata available today.

The first, and probably most relevant cloud-based service, which most readers will have encountered, is Software as a Service or SaaS. This includes design tools, cloud-based accounting software, online games, online movies, etc. The model involves selling someone a license to use software which is hosted on the service provider's computers.

In this crowded space, the book behemoth has created its own line up of services such as Instant Video, book subscription services for its Kindle reading device and more. The potential is so great that it has even been worthwhile to fill its video content channels with original products and TV series which are created in-house. Managed in the typical Amazon data-driven style, series such as *Transparent* were created by analyzing and exploiting its knowledge of consumer tastes. The result? Golden Globes and more.

The next level up would be the Platform as a Service or PaaS offering. PaaS providers are probably less well known to the layman because, to use them, they require you to develop your own software and leverage the platform. But as expected, the big IT companies all compete with products in this space and foremost amongst them is the upstart Amazon.

Your PaaS salesman will look to offer you things like an operating system, database, web server and other necessary tools. You can then build your own programs on this platform and test and run them without the cost or complexity of owning the hardware and underlying software. Examples of such services are Amazon Elastic Beanstalk, Microsoft Windows Azure, which encompasses PaaS and the coming IaaS line, plus other less well-known products.

If you climb the Amazon Elastic Beanstalk, then you will join a host of other software developers who wish to exploit a simple framework for easily deploying their software. You don't have to worry about how much memory or CPUs you need to keep the software running smoothly for your users. You can expand it as needed.

After the SaaSs and PaaSs, there is another level of services which, again, the layman will most likely not recognize. You enter the world of Infrastructure as a Service, or IaaS, which offers computers, either physical or virtual, and other computing resources. It is actually here that Amazon throws most of its hefty weight around.

When Jeff Bezos officially launched his company into the cloud strata, he did so in 2006 with S3. S3 stands for Simple Storage Service and provides customers with a facility in which to store vast quantities of data online. It continues to remain one of the core businesses but it has had several computing compartments added to it.

The infrastructure on which Dropbox launched was another part of Amazon's IaaS, called EC2. It is the Elastic Compute Cloud, and offers flexible computing power, which expands and contracts according to your needs. Another large business which catapulted to success via Amazon's services is the video on demand service provider Netflix.

If we were to go one more level up, you would reach the grand wizard level of Hardware as a Service. Here companies offer services such as leasing or licensing of hardware. There are no platforms, services nor infrastructure; simply computers or other hardware components to hire. In the computing arena, it tends to be the larger hardware players such as IBM, Hewlett-Packard and Fujitsu-Siemens, who offer these services.

In this category, Amazon seems to have chosen to position itself differently. The hardware it wants people to use are its own devices such as the Kindle. Bezos' strategy is different to that of his peers, and when the Kindle Fire tablets were launched in 2012 he said, "We want to make money when people *use* our devices, not when they *buy* our devices."

In between, and around these different classifications, are many more computing models; everything from Desktop as a Service to Backend as a Service. The offerings are becoming more specialized, and as time goes on, we can only expect more categories of the X as a Service type.

Each of the different SaaS, PaaS, IaaS and HaaS are different themes on the same risk lowering innovation technique. Take something which may be expensive to buy or own and offer customers limited access to it for a small fee. It both reduces risk for the customer and for the service provider because it lowers the cost for the consumer and guarantees more users. The interesting part here is that as the cloud service business model spreads to other parts of the economy, we are seeing major innovations occur. It's now mainstream to rent the use of a computer, but what about renting the ability to manufacture?

Open The Floodgates...

As the cloud service model expands, we are seeing this new *modus operandi* infiltrate other areas of the economy not previously aligned with offering their business as a service. It is not hard to imagine that this business model will increasingly change the way in which businesses have traditionally been run—and that could be a good thing.

One area at the cutting edge of innovation and just beginning to embrace the service model, at the time of writing, is manufacturing. In fact, some of the latest and most exciting X as a Service providers are manufacturers. Disrupting typical manufacturing practices with cloud-based services has the potential to have an enormous impact on the entire world; manufacturing comprises around one seventh of the world's entire economic output. It only has industries like government and finance ahead of it.

In order to purchase products from traditional wholesale mass manufacturers, you usually have to purchase a minimum quantity, which can be considerable. Internet based services have the potential to disrupt this, by applying the IT services model to manufacturing, where appropriate. This revolution has already begun, and the goal is to offer people manufacturing services on demand. The trend has arisen out of a push to make manufacturing more flexible. Like many of the other cloud-based software offerings, this one also capitalizes on the Internet's ability to connect people.

The first step in this direction has been made much easier by 3D printing. 3D printing is a way of producing an almost arbitrarily shaped object by building it layer for layer, as explained in the earlier chapter. The marginal cost curve for creating an object in this way is pretty much flat. This means it costs just as much to print your first object as it does to print your thousandth. In addition, a simple 3D plastic printer can be easily purchased by an average consumer—reasonable ones can be had for less than $1,000. However, for high quality results in plastic, or in steel, silver or any other metal, the cost of the printer escalates—enter MaaS.

By pooling numerous high quality 3D printers, able to print in different materials, and opening them up to the public, one creates the most basic form of Manufacturing as a Service. The machines can be hired and used by anyone, making 3D printing affordable for low-volume runs. Users merely need to upload the objects to be printed, have them printed and pay the once-off fee.

This promising business model has not gone unnoticed and in many cities around the word there now exist companies prepared to print your designs on demand. One of the earlier pioneers in this field is a spin out of the Dutch electronics giant, Royal Philips Electronics, called Shapeways.

The idea for Shapeways took place in 2007 within the innovative depths of Philips Electronics. After being spun out of the electronics manufacturer, the startup took the short hop across the Atlantic, landed in New York and set up its printing factories there. Shapeways has since spread to Seattle and has also opened up a new premise in Eindhoven, back home in the Netherlands.

The 3D printing marketplace's success has been twofold. Firstly, it provides a cheap, reliable and rapid prototyping service for people wanting to print their own designs. Secondly, it promotes the creative power of its community by being a platform for them to upload designs. The uploaded designs are then either printed for personal use or sold to others via the Shapeways platform. The entire transaction, from payment to delivery of the printed product, is taken care of by the printing factory.

Their 3D printing on demand service has been a success with the printing of its one-millionth item already in June 2012. In the same year, it built the world's biggest additive manufacturing factory, capable of printing between three and five million creations per year. It now provides 3D design tools to customize designs and has partnered with the multinational toy and board game company, Hasbro Inc., to print toy characters from their shows.

Although impressive, there are limitations to what Shapeways' printers can do. If you want to print your own jewelry or toys, fine. If you want to print a Rolls Royce turbine engine, bad luck. You are not going to have anywhere near the accuracy or consistent outcomes required to print such high quality parts.

Where these services also fall down, when compared to traditional manufacturing, is on cost and speed. As fast as 3D printers are at present, they are still not nearly as fast as more traditional manufacturing processes, such as injection molding.

Injection molding is a process which creates fixed design parts from a mold into which a substance such as molten metal or plastic is injected. For producing tens of thousands of identical objects, this is the perfect process. However, if you need to produce many different designs, a new mold has to be made each time a product with a new shape is manufactured. Despite this, injection molding is still the better method for larger product runs.

In addition, because of the lower manufacturing speed, the production cost of each 3D printed piece is much higher than with traditional manufacturing techniques such as injection molding. So although 3D printing is a good potential service to include under a MaaS umbrella, it does not offer a way of being price competitive for standard, mass produced parts. However, for cheaply prototyping a possible new product, it's ideal.

To achieve a sensible price point for small production runs using older manufacturing techniques, one needs a way to share the costs of production. Websites such as MFG.com are now connecting people wanting manufacturing services with providers of those services. By being able to guarantee the continuous use of equipment for a single factory, it now no longer matters if those orders are coming from one large customer or thousands of Internet connected consumers.

To create an attractive business model for a wide range of factory capacities, one idea is to pool supply and demand on both sides. You can achieve this by having multiple injection molding orders being serviced at once by aggregating many smaller manufacturers hidden behind an Internet connection, a website and a common design requirement. Combining manufacturers and considering their capabilities as a group adds flexibility but keeps costs low at the same time. This is a powerful idea and it is the basis for several large Europe wide projects such as the ManuCloud project.

ManuCloud started in 2010 as a co-operative venture between many countries and large multinationals in Europe, and has committed over five million euros to provide, according to the website,

> "users with the ability to utilize the manufacturing capabilities of configurable, virtualized production networks, based on cloud-enabled, federated factories, supported by a set of software-as-a-service applications."

Its mission is to network resources "down to the shop floor level." Ultimately, the project's ambitious goal is to realize the Manufacturing as a Service dream for industries such as photovoltaic, organic lightning and automotive supply.

Projects such as ManuCloud show the way in how to provide flexibility and cost advantage in production simultaneously. However, there is still an important part of the manufacturing puzzle missing—design. For companies lower down in the supply chain who sell products based on blueprints, design is not a key competitive advantage. But as low cost production moves to other countries, many companies are being forced to add more value to their products and it is design which can help accomplish this.

In recent years, there has been an explosion of competition in this area, with many of the major software firms moving in to try and dominate the market first. Companies such as Autodesk and Sabalcore are now providing cloud-based tools to assist with the design of products which fulfil specific engineering requirements. These tools offer advanced design and simulation capabilities to both develop virtual products and test their physical properties. Users are able to avoid large license fees to test whether their component fulfils the necessary thermal, physical or performance demands required of it. The pay-as-you-go option opens up the use of high-end tools to small and medium sized businesses who simply need to test a new design but have no other design needs.

When we combine both of these services into what is known as cloud-based design manufacturing, then we have truly created a general manufacturing service for the savvy user. The goal is to be cheap, fast and flexible. Entrepreneurs can go beyond mere design and produce their products in one seamless platform. The best part? You pay just for what you use, when you use it! No large upfront investment. A true risk reducer and innovation enabler.

From Manufacturing To Medicine

Given that the "Anything as a Service" movement is disrupting so many traditional business models, where else can we see it going from here? What are the susceptible industries ready to fall victim to this spreading revolution? The next largest target seems to be health care services.

Health care, or medicine, already exists as a service; you don't own a doctor, you merely purchase their services on an as-needed basis. However, what we are currently experiencing is an exporting of more expensive medical procedures to cheaper service-based offerings.

The healthy uprising is being propelled by advanced technology such as robotics, apps and, as always, the Internet. With a slew of Internet doctors waiting to sooth your ailments, the emerging business model now has several of the characteristics of Medicine as a Service: pay-as-you-go, rapid elasticity and ubiquitous access, all hallmarks of the new service paradigm.

Two of the most expensive fields in modern medicine are cancer drugs and specialized surgery. Although the present trend for the former is towards more exotic and expensive medications, a wave of newly developed robots looks to push down the costs for the latter. Small and large surgical machines are set to pave the way for surgeons to operate remotely, and thereby carry out operations on patients in countries on the other side of the globe. A rather dramatic way of opening up new markets.

This new field is called Telesurgery, and is part of a larger movement which has already taken off: Telemedicine. Currently, Telemedicine services are springing up all over the world, with startups trying to connect people with doctors via the Internet or phone. Companies such as NowClinic, Doctor on Demand, American Well and others are striving to disrupt the field of medicine by making possible a diagnosis via video. The service is so well recognized, that even major American health insurance companies such as United Healthcare will currently cover these virtual visits.

But Telemedicine represents only a small part of the technologies helping us monitor and improve our health. Devices installed in our home, observing our behavior and well-being, and connected to doctors waiting to be alerted to a sign of distress could revolutionize how we manage our bodies.

If you think this data collection and management is still in the realm of science fiction, then just look at your wrist. Like one in five Americans, you probably own a so-called health wearable. Devices such as Jawbone, Fitbit and even the Apple Watch contain a set of instruments to measure and gather data on your vital signals and monitor them for your personal pleasure and later viewing. They are conveniently located on your wrist and are switched on 24 hours a day, collecting data about things like your pulse, blood oxygen levels and more. The wearables can be programmed to alert you when your body exhibits dangerous health symptoms; even to call a doctor or ambulance if necessary.

If we want to do more than just monitor ourselves and, for example, require major surgery, the advances on the Telesurgery front will one day reach out and cushion our fall. Telesurgery promises to provide access to a surgeon, wherever you are, as long as the necessary robotic nurses are on-hand. Imagine having your own basic surgery theater at home, ready to help those family members with re-curring surgical needs.

However, these advances didn't occur overnight. What has been achieved so far is the result of the work of many pioneering researchers who have been developing surgical robots for the last 30 years with the aim of performing operations remotely. Our robotic assistants have come a long way since the pioneering operations performed in Vancouver by the so-called Arthrobot.

The Arthrobot was designed to help surgeons with joint restorations. Its inaugural operation was in March 1984, and from there it went on to perform many more surgical procedures over the ensuing 12 months. At the time, it was part of a number of robots developed to assist doctors with various tasks, ranging from handing surgical instruments via voice control, to directly assisting with eye surgery.

As computing power has improved and robotics have become more accurate and reliable, numerous robotic assistants have appeared in surgical theatres worldwide since the late 1990s. They are performing everything from brain surgery to laparoscopy, to difficult forms of microsurgery, and using a wide range of machines to carry out the incision work.

The robots often combine a set of instruments such as cameras or X-ray machines to visualize the patient. The cameras are surrounded by an array of surgical arms poised like an eight-legged praying mantis waiting to perform the highly precise work it has been created for.

Although many of these solutions were revolutionary at their time of introduction, unfortunately they are yet to live up to the promise of reducing the cost of surgery. With some machines costing millions of dollars and also raising the cost of surgical procedures (due to the extra training and maintenance), researchers and doctors have understandably questioned their cost-saving potential. However, as the technology continues to advance, like many industries before it, one expects the return on investment to improve significantly.

In addition, Telesurgery requires more of the one thing which made cloud computing what it is today: bandwidth. For Telesurgery to reach its full potential, the connection and data transfer rates needs to drastically improve in order to realize the dream of operating remotely around the world. Time is critical in an operation and receiving visual clues which are even a few seconds delayed could mean the difference between life and death.

On the other hand, the scope for such services is much greater than simply Medicine as a Service. If a surgeon were to have 30 minutes spare time and would like to carry out a minor procedure, then, via Telesurgery, he or she could begin work immediately from his or her own office PC or Mac.

Programs like, Operation Smiles—through which doctors fly to third world countries to perform reconstructive surgery—would benefit enormously from such technology. Not only could surgeons undertake more operations due to the time saved from flying, but also logistical costs could be saved too, allowing more money to be spent on providing people with lifesaving operations and less on overhead.

Medicine as a Service stands to disrupt the way we experience health care and for many, medical help could potentially be no further away than the smartphone in their pocket.

The Point

The new "Anything as a Service" businesses are presently driving two innovation revolutions. One is by the new companies, which are appearing in almost all segments of the economy, breaking down large expensive items into small financially digestible bits for fiscally constrained entities and so increasing sales. The other innovation is occurring on the receiving end, with startups and smaller companies alike able to purchase new services because they are no longer restricted by big capital or operational outlays. Both of these activities are taking the risk out of innovation for companies by reducing the financial risks for both sellers and buyers.

The thrust of the cloud movement is that before people commit to using a service, they no longer need to worry about large upfront costs or commitments. From being able to test a single use of a service to being able to produce a thousand widgets, it is now possible to test ideas and tools in the real world, and to work in a cheaper and lower risk fashion.

The ability to break down expensive and risky services or products into more affordable parts has become a springboard for innovation. Many large companies have launched their products on the back of other cloud-based services (e.g. Dropbox on Amazon EC2, etc.). The ability to cheaply test multiple ideas on a cloud service to find the one that works is a boon for small cash strapped startups.

Despite all these advances, this promising future has only just begun. New markets and opportunities are arising daily. The integration of rapid elasticity and ubiquitous access to other devices is appearing in the next burgeoning frontier—the Internet of Things.

The Internet of Things is about connecting devices to monitor and provide information on the world around us. When connected to a cloud architecture it can provide you with always accessible information on the current state of affairs of anything from the temperature in your house to the humidity of your plant's soil. By bundling all the necessary technology into a rentable service, it allows the average consumer an easier path into being connected to all the important things in their life.

Another area on the cutting edge of cloud computing which is poised to have a significant real-world impact is the field of boundary-less computing. The true revolution behind this new computing endeavor is the removal of boundaries inside companies via utilizing cloud tools. This means anyone can access data and computing tools from wherever they are and always have the information they need on any part of the company structure. What was typically divided up into corporate silos will now be thrown open, allowing any decision maker to clearly see and quickly analyze the internal workings of each piece of the business puzzle.

For this to occur though, data collection is key, and the act of collection is where employee gadgets and the Internet of Things play an outsized role. The Internet of Things will provide sensors and data collectors throughout a company's complete value creation process and offer managers and employees up-to-date information on every detail, no matter how minor. Combine this with an employee's ability to see this data around the clock from a tablet or smartphone, and the cloud paradigm gets blown in a whole new direction.

There are numerous issues to solve before we reach that point, and we must not forget them. Unanticipated obstacles, such as security vulnerabilities, have prompted a reassessment of what people believed would initially be a simple market to develop and create businesses in. But development has continued and today, if you wish to exploit cloud services, you not only have numerous service providers more than happy to entice you with a nice deal, but the available cloud structures have also multiplied.

For example, what once was simply a private or public cloud system, has now expanded into hybrid cloud versions. Public clouds are the services provided by third parties such as Amazon Web Services. Private clouds are internally created by companies to give cloud-like services in house. The hybrid model internalizes security-sensitive services but externalizes unimportant activities. This has been in response to companies who recognized the need to have tighter security in their IT networks, and who were dissatisfied with the simple public cloud offerings.

Apart from the usual business risks, the thunder-clouds for many XaaS businesses are their razor thin margins. As cloud computing has grown commoditized, more and more companies have dived into the market fray. This has pushed prices down to the point where larger companies are prepared to make a loss on certain simple services in the hope that they can attract customers to other parts of their cloud business. For companies operating only in loss-making markets, there is little hope of survival.

In the meantime, customers of the new services are being offered ways to fundamentally reduce innovation risk. Like other concepts looked at in this book, it has been made possible by the connectivity and functionality of the Internet. The business revolution which simply began with cloud computing has spawned a new way of doing business that has thrown open the doors to low cost and low risk innovation in many sectors and many to come.

Chapter 8
The Bigger Picture

As the world entered the 21st century, the word innovation moved squarely into focus. It became a buzzword, an economic imperative, as well as a billion dollar consulting industry. In the last ten years alone, the number of Google searches for the term *innovate* doubled. So the billion dollar question is, why all the interest?

Over the previous century, technology developments accelerated exponentially, making existing industries redundant, rapidly spawning new ones and leaving many incumbents bankrupt. As things speed up everyone is looking over their shoulder, afraid that the next technological innovation will steamroll them without any warning.

These highly competitive environments are placing companies and people under pressure to innovate ever faster and more efficiently. When the risks of innovation remain high, then you're gambling with your finances—with the odds stacked against you. When firms reduce the risks and costs of innovation though, they guarantee themselves a future. A future in which they can continue to innovate with breakneck pace, but don't break the bank.

For companies and individuals, the argument for reducing their innovation risk is clear. If not being used already, the ideas presented in this book should become a mantra for every organization, every group, and every individual.

But what about the bigger picture? Does reducing innovation risk merely benefit the companies and entrepreneurs or is there something more at stake here?

The answer is revealing and the connections are subtle. Although the data on this topic is limited, the indications are that reducing risk seems to improve society's innovation quality and intensity. This may sound clear, but its consequences are not. The implications are so far reaching and critical that our ability to solve many looming disasters will depend on it, and so it is worth looking at in more detail.

The first assumption people often get wrong is this: just because we cut innovation costs or risk, it does not automatically translate into more innovative activity, even if this is what we hoped for. A reduction in risk may simply mean that more innovations succeed and not that more are embarked upon—*ceteris paribus*, the level of innovative activity may stay the same. But society wants and needs both; more innovative activity and more successful innovations and I believe we are already witnessing this.

The enormous benefits that accrue to society are why reducing risk in the innovation process is so vitally important. I claim that reducing innovation risk has flow-on benefits that are more than just reducing the costs to the risk taker. The hypothesis is that there are at least two significant benefits that accompany a decline in risk.

The first benefit is that more innovations will be successful. This is clear. The second is that the money saved by avoiding failed innovations, and the extra money created by fruitful ventures, will feed back to launch even more ideas and due to the lower risk, more of these will then be successful! This resulting virtuous circle is the real power created by reducing risk in innovation. Companies and individuals naturally want to operate in such a revolutionary environment and the tools presented in this book provide the way to do so.

The logic of the latter benefit is compelling. If innovation becomes less risky, then it means that more innovative activities will succeed. This means that the expected return on such investments will improve and immediately such opportunities will become more attractive as a target for financing. Hence, a virtuous cycle begins simply because of a lower innovation risk.

If we generally assume that these hypotheses do hold, then they have an interesting strategic flip side. It follows that for a fixed sum of money it's better to invest in many smaller innovative activities than in larger ones, if the risks are constant and the outcomes proportional to the investment size. This strategy is not only smarter because it reduces the variance in the overall outcome, but also because it should lead faster into a cycle of more discoveries which succeed sooner and this will then feed back into more innovation.

By now, you are most likely asking, "Is this hypothesis even correct?" Is there any evidence to show that more innovations will occur, and that both the saved and created money will be reinvested to supercharge the whole process? Let us turn to the pharmaceutical industry where we find the first indications of an answer.

The Drug Game

John was devastated. He had just been diagnosed with a malignant melanoma on his leg. Due to its deadly nature, it had to be excised on the same day. However, his worst fears were realized when, within two months after the operation, remnants of the cancer had spread throughout his whole body. The prognosis? One month left to live.

At the time of his diagnosis, there were no existing cancer treatments clinically proven to increase John's survival chances. Given this grim outlook, he did what thousands of others around the world do, every day in fact: he opted to try a new experimental treatment.

As with many experimental drug programs, it started with a trial drug, Brim 3. John was chosen because his tumor tissue demonstrated a specific genetic mutation. He happened to be part of a worldwide study to examine the effects of the drug on hundreds of patients. Luckily for this cancer sufferer, unlike many other risky treatments, there was no cost involved.

By taking this medication, John became a guinea pig for one of the many drugs going through the ubiquitous and stringent government approval process. This approval process is designed to prevent unsafe products from getting in the bloodstreams of the public at large and causing widespread harm and suffering.

It's always a long journey from drug discovery to availability in a pharmacy. For medicines treating heart or brain conditions, the journey can take up to 15 years. It's a path littered with the abandoned husks of failed compounds and billions of dollars.

The whole ordeal begins with the discovery of a molecule which seems to be potentially useful. Computer modelling to identify promising characteristics, combined with vast databases of screened molecules, aid the discovery. This eureka moment can be serendipitous or deliberate, and a bit of both can help an unlikely molecular candidate on its pilgrimage to clinical approval.

After its initial discovery, the new compound must run the gauntlet through at least four major phases of investigation. Even after passing the four phases, many drugs are still continuously monitored by the doctors prescribing it, as well as by the general public. But before this, and before any humans are exposed to a new drug, we turn to our animal friends for help.

The first stage of drug screening involves a group of unsuspecting animals. Normally rats or guinea pigs end up full of chemicals and are then carefully monitored for side effects. For certain drugs, sometimes even pigs are chosen, because their genetic makeup is similar to humans. This allows stronger conclusions to be drawn when necessary. The investigation often involves the testing of both animal sexes, and before mating, to check the effect of the drug on the animal's fertility, as well as for birth defects. At this point, the drug needs to be proven safe in typically larger doses than what people would receive, otherwise its last usage is on these laboratory martyrs.

If the new molecule can jump these hurdles, then it could be on its way to testing on humans, where it then, in many countries, needs to pass three types of trials before it will be deemed acceptable for wider community use. The first step of the trials involve a dozen or so healthy young participants to examine the effect on the human body. If no major side effects are found, then the drug is cleared to move on to the next phase involving the type of patients for which the new medicine was originally designed.

In phase II, the drug is only tested on a limited group of patients suffering from the disease, and they are often people for which little other hope or treatment opportunities exist. These subjects are closely monitored, often in the hospital. Given the novel nature of the product, safe doses need to be established. To achieve this, the drug administration process is repeated several times to find minimum and maximum effective doses.

In phase III, the drug is introduced to a much larger group to study its efficacy in randomized, double-blinded comparisons. Thousands of patients could be exposed to the new compound in this step. Apart from the medical side effects, the cost-effectiveness is also examined, along with a host of other important criteria. In many countries, it is not just the healing ability of a new medicine which is important, but also the cost of that benefit. This can be a major deciding factor as to whether something is approved for prescription, which can also include government subsidies by many national health boards.

It was in the phase III group that John found himself; he was one of the lucky ones. After starting the new drug, his melanomas stopped growing and regressed. To his delight, he enjoyed around five more years of good health, and was even able to play tennis, his favorite game, again. He went overseas, became a ten pin bowling champion and, best of all, was able to see his grandson grow up.

The fairy tale ended, as all do at some point. After years of positive results, the doctors discovered that the cancer had reoccurred in his body. This required swapping to other drugs, but John's kidneys were weak and there were complications with the experimental drugs, plus suffering from renal failure at the same time. A few months later, John went to rest for the last time, at home with his sister caring for him. The melanomas had spread and metastasized to his brain. The extension of life, however, had carried him over into his 80s.

As beneficial as the medication was for John, it wasn't for everyone. With such drugs there are always nasty side effects and this one was no different. This is the reason the drug development process is so rigorous and careful. These phase III tests are the final, necessary obstacle before a drug is released into the community.

If a drug can survive this process and make its way to the wider public, then follow-up studies are also required. These are intended to pick up anything missed in the earlier trials. At this stage, the companies have the most at stake. And this is why the biggest scandals usually occur then.

The Risky Side Of Drug Discovery

The pharmaceutical industry is interesting from many perspectives. It is an industry which spans many disciplines, from life sciences to manufacturing. Large research teams together with efficient manufacturing facilities create a unique blend of high upfront development costs, followed by low production costs for jars of pills or liquids. This unusual corporate combination begins its work with an innovation game that can make or break a company—drug roulette.

The stalwarts of life saving elixirs, it is up to life science companies to develop new products which extend and improve the quality of life for sick people. We rely on them to discover the next blockbuster drugs, Panadol, Aspirin, Nexium, etc. And despite this challenge, the industry, according to the American Congress, "Consistently ranks as one of the most profitable industries in the United States." This profitability is driven by sales of bestseller drugs such as the reflux medicine Nexium, which commands up to $10 billion per year, in the U.S. alone.

With sales in the billions and growth in many markets, what more could Big Pharma need?

How about winning our trust back?

The last 20 years have seen an unprecedented series of bad headlines and court cases for the world's pharmaceutical industry. What's worse is that the list of unsafe and dangerous drugs implicates the who's who of the largest and most respected companies. From GlaxoSmithKline to AstraZeneca, each has released a powerful drug to solve one health problem, but created an even worse one in its place.

A high profile example of this situation was the drug Avandia. Introduced to combat type 2 diabetes, this medicine became a major cash cow for GlaxoSmithKline. At its peak, the new compound was earning its creators between two and three billion dollars a year. It seemed to be a hit, assisting people with diabetes around the world. The medicine could have been a game-changer, if it weren't for one nasty side effect.

Avandia was killing people.

The American Food and Drug Administration pulled the plug on its unconditional support of the drug during 2007 due to an estimated 80,000 people suffering from heart attacks or strokes as a result of taking Avandia. In 2010, the use of the drug was severely restricted, with only a small group of special patients allowed to receive treatment. In the same year, Avandia was banned in Europe. To underline the severity of the situation, the venerable New England Journal of Medicine published research showing, "The medicine increased heart attack risk by a whopping 43 percent."

GlaxoSmithKline admitted its guilt in 2012 and paid out over $3 billion in fines for, according to Time, "failing to report drug safety information." It was the largest settlement paid by a pharmaceutical company at the time. This dodgy drug joined a list of other misrepresented and banned medicines from the company's portfolio such as Paxil and Wellbutrin. Why did this happen?

Developing new drugs is a highly risky business. It's an industry where the estimated costs of new drug development are now around one billion dollars. This type of risk can mean that after a couple of expensive failures, you're broke—even for large multinationals. Add to this the difficulties in even finding useful drugs, followed by patent expiry dates, and you can understand the pressure companies are under, which too often leads them to making illegal and amoral decisions.

Apart from a desperate need to acquire better moral compasses, Big Pharma is an industry in dire need of help due to "Escalating costs and the declining number of launches," as stated in an article in the world-renowned Nature Reviews Journal. The authors continue,

> "[W]ithout a substantial increase in R&D productivity, the pharmaceutical industry's survival (let alone its continued growth prospects), at least in its current form, is in great jeopardy."

Since 1990, the total investment in research by the drug industry has tripled in real terms. However, the number of innovative new drugs approved by the American Food and Drug Administration has remained relatively flat. The indication is then, that companies are getting fewer results for more research spending. It is also an extremely research intensive industry, investing five times more money into research and development than the average U.S. manufacturing firm.

Given this level of financial risk, much research has been conducted to try and understand the situation. All aspects of the drug delivery model have been analyzed; from research and development outputs, to works in progress, technical success probabilities, cycle times and more. As a result, we know more about the economic and innovation risks in the pharmaceutical industry than in almost any other industry. The rich data available provides one with a fertile testing ground for all sorts of hypotheses, especially innovation ones.

Amongst the details of the research effort devoted to understanding the nature and process of drug discovery, one finds evidence to support the hypothesis that reducing innovation risks leads to more innovation and a resulting virtuous innovation cycle. The data points to the conclusion that when lower risk leads to more success, more money is invested back into research and development—the virtuous cycle we were hoping for.

As to the rationale behind this, research by the Congressional Budget Office of the United States shows that,

> "[T]he relative stability of the relationship between pharmaceutical R&D and sales revenue suggests that firms find it most profitable to invest any additional dollar of sales revenue in their own drug research."

So for each dollar saved by using new, less risky innovation techniques, we see that not only will firms create more innovations, but they will reinvest the saved money into further research and development. This is exactly the desired behavior.

The authors continue by writing that lowering the barrier and costs for new drug application reviews "induced firms to complete, and to prepare approval applications for, their late-stage development projects more quickly." So by reducing the red tape for approvals, i.e. the risk associated with the drug development process, innovation progresses faster. Combined with the money and time saved with such streamlined processes, the results will hopefully be what we want from the industry: more blockbuster drugs.

Apart from the findings of the United States congressional publications, models presented in the Nature Reviews Drug Discovery study showed that when innovation risks are reduced, a disproportionate gain in the number of successful drugs is realized. This virtuous cycle arises because less money is being spent on riskier drugs, which means that funds can then be reinvested and used to discover and test new drugs.

So, taking the example of the pharmaceutical industry, the evidence suggests that when we lower innovation risks, we should get exactly the desired outcomes of more successful innovations and more innovation occurring. Although we are yet to see this benefit (as stated earlier research inputs are rising and outputs remain flat), the industry appears to present extremely fertile ground to start employing risk reducing innovation tools to start a radical virtuous cycle and wrench the pharmaceutical industry out of its current quagmire.

These compelling benefits are some of the strongest reasons for society to pursue, use and develop more "game-changing" tools which are able to reduce innovation risk, and these are precisely the tools that are presented in this book. Hopefully, by mastering the techniques presented here, Big Pharma will be less inclined to peddle defective drugs and be able to sell more blockbusters without harmful side effects. A win-win situation for all.

The Point

Does this lesson then apply to you? The answer is a resounding yes. Apart from simply improving your own chances of succeeding, more people will innovate more often, and will also succeed more often, and this will most likely lead to a higher level of innovation and growth for all. The techniques you have read about here empower you and society to do exactly that.

Have a large job to do, or to finance? Turn to the crowd. Don't have the money to buy the machines needed to realize your dreams? Join a hackerspace. Need more opinions and ideas in your company? Open up your innovation portal to the world. Want to know how to better invest in and manage your innovations? Then look at the latest on behavioral innovation to make sure you are not over-investing in yesterday's technology. Does cutting edge knowledge drive your business? Then make sure you support and use your open access options. Do you have an expensive service or need to spread out your operational costs over time? Then look to the X as a Service model to grow your business.

A core part of what's driving the majority of the new low risk innovation techniques has been advances in technology, and in particular digital technologies. In most cases, this technology was initially the Internet, but in all cases it was based on powerful, low-cost ways to experiment with innovative ideas. As the professor of management at the MIT Sloan School of Management, Erik Brynjolfsson states,

> "Technology is transforming innovation at its core, allowing companies to test new ideas at ... prices that were unimaginable even a decade ago."

By putting these technologies and techniques in your innovation toolbox, and mastering how and when to use them, you will discover a new paradigm of low risk and cheap innovation. In fact, the patent master Thomas Edison once said that he had never failed, but had found 10,000 ideas which didn't work. If the light bulb inventor had not managed risk so well, and instead poured all his resources into every new idea, then he would have likely gone broke and given up after the first few failed inventions.

The bigger picture is that low-risk innovation begets more innovation, which succeeds for the reasons explained in this chapter. But, the upshot is even bigger than this. As we saw in the introduction, if innovation becomes a low-risk game, more people will get involved because the risk level will drop below their participation threshold. The message now becomes clear, if innovation drives what you do, then making it low risk will massively catalyze your success. This is the real answer to the billion dollar question we posed at the start.

We have seen a snapshot of the available tools to lower risk in innovation. I am sure there are others you will find that are also useful and grant you an innovative edge. But more than just deploying these tools, I hope the topics discussed here will inspire you to create your own ways to reduce the risk of innovation. Whether employing new methods, new technologies, or whatever else it takes, you'll become a true "innovative innovator."

Acknowledgments

First and foremost, I would like to thank my wife and family for their support and patience during this adventure. There were many points along the way where I asked myself, what the hell I was doing but with the support of my family I made it. I also am grateful for my partner's editorial assistance and keeping my big picture focus aligned with the goals that I wanted to achieve in this book.

I would also like to thank my parents for their help in many ways, not in the least with editing and improving the quality and content of the final product. Without that, many half-formed ideas would have remained so on the pages, half-formed and half-comprehensible. In particular, I'd like to thank my mother who took the editing process to another level by completely transforming a very rough text into something intelligible.

It is also important for me to thank both my diligent and professional editors Hayley Ward and Richard Peters. The final touches you both put on this book added that extra shine to turn it into something special. I'd especially like to point out Richard Peters for his commitment to taking this book to the next level and improving it above and beyond what was expected. Thanks for that! Any mistakes not found by the two editors are of course my own.

I'd also like to thank Saba Tekle for helping me along this journey and also being a fantastic publisher and mentor. I am extremely grateful that there are people such as yourself in this world who help little known authors break out into the literary world.

Also a big thank you to all the people who read my manuscript and provided feedback such as Amir Elion, Christine Thong, Anita Kocsis, Steve Glaveski, Mari Anixter, Rowan Gibson, David Fradin and many more.

Notes and References

Chapter 1—Introduction

"Research, however, has shown there are many factors in place," see the paper, Tsui A. B. *Reticence and anxiety in second language learning.* (1996). Voices from the language classroom, pp. 145-167. In this paper the author describes issues that students have when learning a second language and their fears preventing them from being able to participate as much as they'd like.

"Research has shown that given an equal chance of winning or losing, we consider the chance of failure to far outweigh the equivalent possible gains", these comments refer to Kahneman D. and Tversky A. Prospect theory: An analysis of decision under risk. (1979). *Econometrica: Journal of the Econometric Society*, pp 263-291. This groundbreaking paper which has been cited almost 40,000 times at the time of writing, laid the groundwork for a descriptive utility theory unlike the von Neumann and Morgenstern normative framework. I.e. it tried to describe how real people behave unlike how people should behave if they were perfectly rational.

"The pair who began that particular fast food chain came up with the idea after visiting one of the McDonald Brothers' original stores," see the book by Jakle J. A. and

235

Sculle K. A. *Fast food: Roadside restaurants in the automobile age.* (2002). JHU Press. They describe the original founders of Burger King as wanting to copy the success of McDonalds.

"it has been shown that people's innovation threshold varies wildly", see Miller P. and Wedell-Wedellsborg T. *Innovation as usual: How to help your people bring great ideas to life.* (2013). Harvard Business Review Press. They describe some work I cite later.

"[C]reativity is in large part a decision that anyone can make but that few people actually do make because they find the costs to be too high", this quote is from Sternberg R. J. *The nature of creativity.* (2006). Creativity Research Journal, 18(1), 87-98.

"research has shown that we tend to be more risk averse when gambling for a gain than when trying to avoid the same loss", see the aforementioned paper by Kahneman D. and Tversky A. Prospect theory: An analysis of decision under risk. (1979). *Econometrica: Journal of the Econometric Society*, pp 263-291

"Scientist have even been able to demonstrate this type of behavior in rats", see Zalocusky K. A., Ramakrishnan C., Lerner T. N., Davidson T. J., Knutson B. and Deisseroth K., *Nucleus accumbens D2R cells signal prior outcomes and control risky decision-making*, 23 March 2016, Nature, doi:10.1038/nature17400

"It has been shown that as few as 7% of all innovators successfully commercialize their inventions", see Dodgson M., Gann D. M. and Salter A. *The management of*

technological innovation: strategy and practice. (2008). OUP Oxford. See, for example, Box 3.13.

"Other studies have shown that 4% of innovation initiatives achieve their internally defined success criteria. Only 12% of research and development projects even return their capital cost", see Franklin C. *Why innovation fails: Hard-won lessons for business.* (2005). Spiro Press.

"leads to increased profits, market value, survival in the market jungle and much more" and *"During this phase Edison tested horsehair, cork, rubber, grass fibers, wrapping paper, fishing line, silk and even the beard hair from his laboratory workers"*, see the aforementioned book by Dodgson M., Salter A. and Gann D. *Think, play, do.* (2005). Oxford Univ. Press. In particular, see pages 15, 16 and 18.

"When two young budding entrepreneurs tried to sell their business to Excite for $750,000, the CEO of Excite rejected the offer thinking the company was not worth the money", see the article by Siegler M. G. *When Google Wanted To Sell To Excite For Under $1 Million—And They Passed.* (29 September 2010). TechCrunch. Retrieved February 21, 2016, from http://techcrunch.com/2010/09/29/google-excite/

Chapter 2—Share The Risk And Increase The Gain

"From early women's liberation", see the article by O'Malley, S. (May, 2008). *The Importance of the Bicycle to the Early Womens Liberation Movement.*

CRANKEDMAG. Retrieved February 21, 2016, from https://crankedmag.wordpress.com/issues/issue-4/the-importance-of-the-bicycle-to-the-early-womens-liberation-movement/

"The effect Wikipedia has had on the lives of net citizens is huge", the information following this sentence is based on largely on Wikipedia. (2016, February 19). In *Wikipedia, The Free Encyclopedia*. Retrieved February 21, 2016, from https://en.wikipedia.org/w/index.php?title=Wikipedia&oldid=705741476

"The "wiki" is one of the top ten most visited websites in the world", see List of most popular websites. (2016, February 14). In *Wikipedia, The Free Encyclopedia*. Retrieved February 21, 2016, from https://en.wikipedia.org/w/index.php?title=List_of_most_popular_websites&oldid=704919599

"Wikipedia entries are one and a half times more likely to refer to open access articles than closed access articles", see Emerging Technology from the arXiv. *Why Wikipedia + Open Access = Revolution.* (2 July 2015). Technology Review. Retrieved February 21, 2016, from http://www.technologyreview.com/view/539001/why-wikipedia-open-access-revolution/

"An early example of crowdsourcing is the definitive Oxford English Dictionary", see Lanxon, N. *How the Oxford English Dictionary started out like Wikipedia.* (13 January 2011). Wired. Retrieved February 21, 2016, http://www.wired.co.uk/news/archive/2011-01/13/the-oxford-english-wiktionary.

"A quick search revealed more than 35 such platforms operating at the time of writing", see Comparison of crowdfunding services. (2016, February 19). In *Wikipedia, The Free Encyclopedia*. Retrieved February 21, 2016, from https://en.wikipedia.org/w/index.php?title=Comparison_of_crowdfunding_services&oldid=705856222

"Located in Brooklyn, New York", see Bergl S., Cattel J., and Kratochwill L. *True To Its Roots: Why Kickstarter Won't Sell.* (2013, March 18). Fast Company. Retrieved February 25, 2016, from http://www.fastcompany.com/3006694/where-are-they-now/true-to-its-roots-why-kickstarter-wont-sell. The quotes in the next paragraph are also from the same article.

"To date, Kickstarter has achieved the highest funded projects for any crowdfunding platform", see Comparison of crowdfunding services. (2016, February 23). In *Wikipedia, The Free Encyclopedia*. Retrieved February 25, 2016, from https://en.wikipedia.org/w/index.php?title=Comparison_of_crowdfunding_services&oldid=706538542

"In fact, Games raised two out of every ten dollars pledged on Kickstarter in 2013", see Stats. Kickstarter. Retrieved April 14, 2017 from https://www.kickstarter.com/help/stats

"about 10% of the films accepted by the 2012 Sundance, Tribeca, and South by Southwest Film Festivals", Kickstarter Blog. Kickstarter. Retrieved April 10, 2016, from https://www.kickstarter.com/blog/100-million-pledged-to-independent-film.

The facts about Indiegogo's background are based, in part, on the information found at, Indiegogo. (2016, January 8). In *Wikipedia, The Free Encyclopedia*. Retrieved February 25, 2016, from https://en.wikipedia.org/w/index.php?title=Indiegogo&oldid=698878185

"Kickstarter and Indiegogo have launched somewhere in the order of 300,000 projects at the time of writing, with just under 100,000 of those on Kickstarter reaching their goals and very likely with a similar number on Indiegogo", see Yoda. *Kickstarter vs Indiegogo – By The Numbers*. (2013, September 4). StartupsFM. Retrieved February 25, 2016, from http://startups.fm/2013/09/04/kickstarter-vs-indiegogo-by-the-numbers.html **and** Kickstarter. *Stats*. Kickstarter. Retrieved February 25, 2016, from https://www.kickstarter.com/help/stats?ref=press

"these projects have raised somewhere in the order of $3 billion dollars to back ideas from around the entire world", see Stats. Kickstarter. Retrieved April 14, 2017 from https://www.kickstarter.com/help/stats

"You are four times more likely", see madd. The ABC of BACs. Madd. Retrieved February 25, 2016, from http://www.madd.ca/media/docs/ABCs%20_of_BACs_FINALdoc.pdf

"This is about the same as downing two standard glasses of wine then jumping behind the wheel", see *What is a standard drink?* Retrieved February 25, 2016, from http://www.druginfo.adf.org.au/fact-sheets/what-is-a-standard-drink-web-fact-sheet

"*quarter of all the land area being at or below sea level*", see Worldatlas. Netherlands Geography. Retrieved March 17, 2016, from http://www.worldat-las.com/webimage/countrys/europe/nether-lands/nlland.htm

"*However, as he was of Canadian origin and not Dutch*", the text in the following paragraphs comes from Smartwatch FM. *Smartwatch FM interview with Eric Migicovsky.* (2013, July 29). Technology.FM. Retrieved March 17, 2016, from http://www.technology.fm/leaders/ericmigicovsky

"*was a little Arduino circuit board and a Nokia phone screen*", see Dingman S. *Why Vancouver-born Pebble smartwatch founder isn't afraid of Apple and Google.* (2014, September 8). The Globe And Mail. Retrieved March 17, 2016, from http://www.theglobe-andmail.com/report-on-business/careers/careers-leader-ship/high-tech-tailor-measures-up-in-the-silicon-val-ley/article20378020/?page=all

"*was by Microsoft back in 2003*", see Carnoy D. *Microsoft watch keeps up with the times.* (2006, June 14). Cnet. Retrieved March 17, 2016, from http://news.cnet.com/Microsoft-watch-keeps-up-with-the-times/2100-1041_3-6083432.html

"*Eric's next prototype was a product called InPulse*", the text in the following paragraphs comes from Sawh M. *TR Talks: Pebble on the Apple Watch, Android Wear and more.* (2015, January 2). Trusted Reviews. Retrieved

March 17, 2016, from http://www.trustedre-
views.com/opinions/pebble-smartwatch-interview-with-
myriam-joire#4mj5LFGt7pGM4tWm.99

"It wasn't waterproof, had a seven hour battery life" as
well as the text following *"InPulse was also a complete
failure and almost cost Eric his company"*, see Dingman
S. *Why Vancouver-born Pebble smartwatch founder
isn't afraid of Apple and Google.* (2014, September 8).
The Globe And Mail. Retrieved March 17, 2016, from
http://www.theglobeandmail.com/report-on-busi-
ness/careers/careers-leadership/high-tech-tailor-
measures-up-in-the-silicon-valley/arti-
cle20378020/?page=all

"In spite of this failure, the team learnt a lot", the text in
the following paragraphs comes from Sawh M. *TR Talks:
Pebble on the Apple Watch, Android Wear and more.*
(2015, January 2). Trusted Reviews. Retrieved March 17,
2016, from http://www.trustedreviews.com/opin-
ions/pebble-smartwatch-interview-with-myriam-
joire#4mj5LFGt7pGM4tWm.99

"Within only two hours of going live", the text in the following paragraphs comes from Netburn, D. *Pebble
smartwatch raises $4.7 million on Kickstarter funding
site.* (April 18, 2012). LAtimes.com. Retrieved March 17,
2016, from http://www.latimes.com/business/technol-
ogy/la-fi-tn-pebble-smart-watch-kickstarter-
20120418,0,1769291.story

*"From helping fund top students unable to pay for their
college education"*, see Jacobs P. *The Inspiring Story of
How A Boston University Student Raised $US5,000 In A*

Day To Stay In College. (2013, October 24). Business Insider Australia. Retrieved March 17, 2016, from http://www.businessinsider.com.au/inspiring-story-boston-university-student-raised-5000-stay-college-2013-10

"*or paying unaffordable medical bills*", see gofundme. *Success Stories*. Retrieved March 17, 2016, from http://www.gofundme.com/success/
"*However, crowdfunding can have a negative side for the investor*", the text in the following paragraphs is based on facts from Kickstarter. (2016, February 19). In *Wikipedia, The Free Encyclopedia*. Retrieved February 21, 2016, from https://en.wikipedia.org/w/index.php?title=Kickstarter&oldid=705700896

"*who refused to repay the promised funds after his board game project failed*", see Fingas J. *FTC starts cracking down on crowdfunding fraud*. (2015, November 6). Engadget. Retrieved March 17, 2016, from http://www.engadget.com/2015/06/11/ftc-cracks-down-on-crowdfunding-fraud/

"*There have been cases*", see Kickstarter. (2016, February 19). In *Wikipedia, The Free Encyclopedia*. Retrieved February 21, 2016, from https://en.wikipedia.org/w/index.php?title=Kickstarter&oldid=705700896

The quote "*it could create more risk than protection*" is from Greenberg A. *Privacy Router Anonabox Gets $600K In Crowdfunding – And Huge Backlash*. (2014, October 16). Wired. Retrieved March 17, 2016, from http://www.wired.com/2014/10/anonabox-backlash/

"*such as a crowdfunded potato salad*", see Kickstarter. *Potato Salad*. Retrieved March 17, 2016, from https://www.kickstarter.com/projects/324283889/potato-salad

"*Books have even been written to help people "crowdfund" themselves*", see Johnson S. *Crowdfund 2.0 Launch Formula: Your Ultimate Guide to Raising Money For Any Financial Need or Project You Have by Helping Others!* (2015). Amazon Digital Services LLC.

"*is estimated to save consumers up to $60 billion per year*", see Rothwell R. *Creating wealth with free software*. (2008, September 5). Free Software Magazine. Retrieved March 17, 2016, from http://www.freesoftwaremagazine.com/articles/creating_wealth_free_software

"*or twice what is needed to feed the world's undernourished*", see StopTheHunger.com. Retrieved March 17, 2016, from http://www.stopthehunger.com/

"*the users have the freedom to run, copy, distribute, study, change and improve the software*", see GNU Operating System. *The GNU Manifesto*. Retrieved March 17, 2016, from http://www.gnu.org/gnu/manifesto.en.html#mission-statement

"*squarely in every programmers' focus*", see Kelty C. M. *Two Bits*. (2008). Duke University Press.

"*most popular and liberal programming licenses available*", for more information see MIT License. (2016, March 5). In *Wikipedia, The Free Encyclopedia*. Retrieved

March 17, 2016, from https://en.wikipedia.org/w/index.php?title=MIT_License&oldid=708402716

"There are more than 1400 options", see Pham A. T., Weinstein M. B. and Ryerson J. L. *Easy as ABC: Categorizing Open Source Licenses.* www.IPO.org. Retrieved 14 April, 2017.

"which has more restrictive conditions on code sharing and sees open source as "amoral"", see GNU Operating System. *The GNU Manifesto.* Retrieved March 17, 2016, from http://www.gnu.org/gnu/manifesto.en.html#mission-statement

"Hence they are enforceable under current copyright law", see Shiels M. *Legal milestone for open source.* (2008, August 14). BBC News. Retrieved March 17, 2016, from http://news.bbc.co.uk/1/hi/technology/7561943.stm

"although larger projects such as SourceXchange and Eazel have also failed", see Sharma S., Sugumaran V. and Rajagopalan B. *A framework for creating hybrid-open source software communities (PDF).* (2002). Info Systems Journal 12: 7–25. doi:10.1046/j.1365-2575.2002.00116.x

"minor changes can demand the attention of two independent programmers", see Tripp A. *Classpath hackers frustrated with slow OpenJDK process.* (2007, July 16). Javalobby. Retrieved March 17, 2016, from http://www.javalobby.org/java/forums/t98834.html

"which may be harder for them to find in closed source software", see Gallivan M. J. *Striking a Balance Between*

Trust and Control in a Virtual Organization: A Content Analysis of Open Source Software Case Studies. (2001). *Info Systems Journal* 11 (4): 277–304. doi:10.1111/j.1365-2575.2001.00108.x

"*the response by some multinational companies to their known security flaws*", this statement refers to the Lenovo Adware hack which was appallingly managed by the company. For more details see McCormick R. *Security researchers found another 'massive security risk' in Lenovo computers.* (2015, May 6). The Verge. Retrieved March 17, 2016, from http://www.theverge.com/2015/5/6/8557881/security-researchers-found-another-massive-security-risk-in-lenovo

"*allowed the company to stay afloat*", see Eadicicco L. *THE STORY OF ANDROID: How a flailing startup became the world's biggest computing platform.* (2015, March 29). Business Insider Australia. Retrieved March 18, 2016, from http://www.businessinsider.com.au/how-android-was-created-2015-3?r=US&IR=T

"*80% of all smartphones shipped around the world had a version of Android on them*", see Kerr D. *Android dominates 81 percent of world smartphone market.* (2013, November 12). Cnet. Retrieved March 17, 2016, from http://www.cnet.com/news/android-dominates-81-percent-of-world-smartphone-market/

"*Android has been a massive financial success for Google*", see Judge S. *Android a Financial Success for Google.* (2011, February 9). Mobile Phone Development. Retrieved March 17, 2016, from http://www.mobilephonedevelopment.com/archives/1204

"Mozilla Firefox is the second most used desktop web browser, beaten only by Google's Chrome browser", see *StatCounter Global Stats—Browser, OS, Search Engine including Mobile Usage Share.* statcounter.com. Retrieved March 24, 2016, from http://gs.statcounter.com/#desktop-browser-ww-monthly-201508-201602

"Netscape Navigator had almost 80% of web traffic accessing the internet via its software", see Gromov G. *Roads and Crossroads of the Internet History.* NetValley.com. Retrieved March 24, 2016, from http://www.netvalley.com/cgi-bin/intval/net_history.pl?chapter=4

"Even the BBC has joined their ranks with its own tiny computer, the Micro Bit", see Wakefield J. *BBC says Micro Bit rollout will be delayed.* (2015, September 17). BBC News. Retrieved March 17, 2016, from http://www.bbc.com/news/technology-34281688

The quote *"a working plastic gun that could be downloaded and reproduced by anybody with a 3D printer"* comes from Poeter D. *Could a 'Printable Gun' Change the World?* (2012, August 24). PC Mag. Retrieved March 17, 2016, from http://www.pcmag.com/article2/0,2817,2408899,00.asp.

"public safety threat", see Timms A. *The future of 3D printing might be scarier than you thought.* (March 29, 2013). *Blouin News.* Retrieved March 17, 2016, from http://blogs.blouinnews.com/blouinbeattechnology/2013/03/29/the-future-of-3d-printing-might-be-scarier-than-you-thought/

"*to Armageddon*", see Zaleski A. *Cody Wilson Wants to Destroy Your World.* (2015, March 11). BackChannel. Retrieved March 17, 2016, from https://medium.com/back-channel/cody-wilson-wants-to-destroy-your-world-ad121c8b0a6

"*introduce legislation to ban making such weapons*", see Franzen C. *NY Congressman Introducing Ban on 3D-Printed High Capacity Gun Magazines.* (2013, January 16). TPM. Retrieved March 17, 2016, from http://talkingpointsmemo.com/livewire/ny-congressman-introducing-ban-on-3d-printed-high-capacity-gun-magazines

"*questioned the reliability of such parts*", see Greenberg A. *I made an untraceable AR-15 'Ghost Gun' in my office – and it was easy.* (2015, March 6). Wired. Retrieved March 17, 2016, from http://www.wired.com/2015/06/i-made-an-untraceable-ar-15-ghost-gun/

"*Finally, open source biotechnology*", the text in the following paragraphs is based on Open Source Biotechnology. (2016, March 3). In *P2Pfoundation*. Retrieved March 17, 2016, from http://p2pfoundation.net/Open_Source_Biotechnology

"*An unassuming vial with a barely visible speck of goo*", the text in the following paragraphs is based on Schloendorn J. *Chapter 2. Open Source Biotech Consumables.* (2013). O'Reilly Media. Retrieved March 17, 2016, from http://chimera.labs.oreilly.com/books/1234000002036/ch02.html#_pricking_the_bubble

"*Numerous writers have complained about the whim of the crowd*", see Cooney M. *The whim of the crowd*. (2010, November 10). Idealog. Retrieved March 17, 2016, from http://idealog.co.nz/venture/2010/11/whim-crowd

Chapter 3—The Latest Tools And Space—For A Dime

"*Wurundjeri, Boonwurrung, Taungurong, Dja Wurrung and Wathaurung people of aboriginal lineage*", see *Aboriginal Melbourne*. City of Melbourne. Retrieved March 10, 2016, from http://www.thatsmelbourne.com.au/Placestogo/indigenous/Pages/indigenous.aspx

"*led to the city becoming the second most culturally diverse place*", see Glenn. 2011 *Census – Where are the most multicultural communities?* (2012, August 14). Id. Retrieved March 11, 2016 from http://blog.id.com.au/2012/population/australian-census-2011/2011-census-multicultural-communities/

"*second most multicultural country in the world: Australia*", see Griffiths M. *Australia second most multicultural country*. (2010, November 17). ABC News. Retrieved March 10, 2016, from http://www.abc.net.au/news/2010-11-17/australia-second-most-multicultural-country/2339884

"*largest Greek population outside of Europe after Athens and Thessaloniki*", see France-Presse A. *Australian heads to Greece to help pensioner pictured crying outside bank*. (2015, July 8). The Guardian. Retrieved March 10,

2016 from http://www.theguard-
ian.com/world/2015/jul/08/australian-heads-to-greece-
to-help-pensioner-pictured-crying-outside-bank

The information following *"walk towards the historic
Flinders Station from the city"* is based on Flinders Street
railway station. (2016, February 25). In *Wikipedia, The
Free Encyclopedia*. Retrieved March 12, 2016, from
https://en.wikipedia.org/w/index.php?title=Flin-
ders_Street_railway_station&oldid=706748281

The information following *"The Connected Community
Hackerspace Melbourne was founded in 2009 by Andy
Gelme"* is based on http://hackerspaces.org/wiki/Con-
nected_Community. Retrieved March 12, 2016.

The text following *"Not only does Andy have the chassis
for the first Cray in Australia"* is based on Brewster S.
The search for the lost Cray supercomputer OS. (2014,
January 14). Gigaom. Retrieved March 12, 2016 from
https://gigaom.com/2014/01/14/the-search-for-the-lost-
cray-supercomputer-os/

*"Makerspaces tend to be more about making things with
off-the-shelf products"*, see Cavalcanti G. *Is it a Hackerspace, Makerspace, Techshop, or FabLab?* (2013, May
22). Make:. Retrieved March 11, 2016, from
http://makezine.com/2013/05/22/the-difference-be-
tween-hackerspaces-makerspaces-techshops-and-fab-
labs/

The quote *"to exploit new computer technologies to meet
the needs of military command and control against nuclear threats"* comes from Lukasik S. J. *Why the Arpanet*

Was Built. IEEE Annals of the History of Computing 33 no. 3 (2011): 4-20.
http://dx.doi.org/10.1109/MAHC.2010.11

"made evident the loopholes in fingerprint security systems", see *German Defense Minister von der Leyen's fingerprint copied by Chaos Computer Club.* (2014, December 28). DW. Retrieved March 11, 2016, from
http://www.dw.com/en/german-defense-minister-von-der-leyens-fingerprint-copied-by-chaos-computer-club/a-18154832

"As of 2012, there is an estimated 700 to 1,200 active such community workspaces all over the world", see Hackerspaces.org. *List of Hacker Spaces.* Hackerspaces. Retrieved April 14, 2017.

Much of the information in the paragraphs following *"A cheap mill or lathe can cost roughly $5,000—$20,000"* is based on Hatch, M. (2013). *The maker movement manifesto: rules for innovation in the new world of crafters, hackers, and tinkerers.* McGraw Hill Professional.

The patent *"Apparatus and Method for creating three-dimension objects"* is Crump, S. S. (1992). *U.S. Patent No. 5,121,329.* Washington, DC: U.S. Patent and Trademark Office.

"In fact, over 34,000 machines were crowdfunded over the course of four years", see *On Trend: Successfully Crowd-Funded 3D Printing Projects.* (2014, November 24). Spark. Retreived on March 11, 2016, from

http://spark.autodesk.com/blog/trend-successfully-crowd-funded-3d-printing-projects

The text following *"By the 1920s, adding layers of metal as a manufacturing process"* is based on Colegrove P. *High deposition rate high quality metal additive manufacture using wire + arc technology.* Retrieved March 11, 2016, from https://xyzist.com/wp-content/uploads/2013/12/Paul-Colegrove-Cranfield-Additive-manufacturing.pdf

"Otto John Munz's 1951 invention", see John, M. O. (1956). *U.S. Patent No. 2,775,758.* Washington, DC: U.S. Patent and Trademark Office.

"produce complex jet engine parts was filed by a Ross F. Housholder in 1979", see Housholder, R. F. (1981). *U.S. Patent No. 4,247,508.* Washington, DC: U.S. Patent and Trademark Office.

"such as when the Stratasys patent's term ended", see Robinson J. *Expiring patents were supposed to boost the 3D printing market.* They haven't. (2014, June 6). Pando. Retrieved March 11, 2016, from https://pando.com/2014/06/06/expiring-patents-were-supposed-to-boost-the-3d-printing-market-they-havent/

"or even fundamentally new wing designs for unprecedented aerodynamic efficiency, are now possible", see Bullis K. *Additive Manufacturing Is Reshaping Aviation.* (2015, February 6). MIT Technology Review. Retrieved March 11, 2016, from http://www.technologyreview.com/news/534726/additive-manufacturing-is-reshaping-aviation/

"If adopted by all airlines, this would be the equivalent fuel needed to run all the cars in America for half a year", see Rockstroh T., Abbott D,, Hix K. and Mook J. *Additive manufacturing at GE Aviation.* (2013, November 26). Industrial Laser Solutions. Retrieved March 11, 2016, from http://www.industrial-lasers.com/articles/print/volume-28/issue-6/features/additive-manufacturing-at-ge-aviation.html

The quote *"3D printer that prints itself"* is from RepRap. (2016, January 19). *RepRap Wiki.* Retrieved March 11, 2016, from http://reprap.org/mediawiki/index.php?title=RepRap&oldid=165471

"most widely used among the global members of the Maker Community", see RepRap. (2016, January 19). *RepRap Wiki.* Retrieved March 11, 2016, from http://reprap.org/mediawiki/index.php?title=RepRap&oldid=165471

"owning a RepRap is an economically sound investment", see Wittbrodt, B. T., Glover, A. G., Laureto, J., Anzalone, G. C., Oppliger, D., Irwin, J. L., & Pearce, J. M. (2013). *Life-cycle economic analysis of distributed manufacturing with open-source 3-D printers. Mechatronics, 23*(6), 713-726.

The quote *"[RepRap] has been called the invention that will bring down global capitalism, start a second industrial revolution and save the environment..."* is from Randerson J. *Put your feet up, Santa, the Christmas machine has arrived.* (2006, November 25). The Guardian. Retrieved March 11, 2016, from http://www.guardian.co.uk/christmas2006/story/0,,1956793,00.html

The text following *"Future body parts on order seem to be on the horizon"* is based on Ledford H. *The printed organs coming to a body near you.* (2015, April 15). Nature News. Retrieved March 11, 2016, from http://www.nature.com/news/the-printed-organs-coming-to-a-body-near-you-1.17320

The text following *"One of the more common condiments being extruded from a printer head is chocolate"* is based on Porter K., Phipps J., Szepkouski A., and Abidi S. *3D Opportunity Serves It Up.* (2015, June 18). Deloitte University Press. Retrieved March 11, 2016, from http://dupress.com/articles/3d-printing-in-the-food-industry/

"researchers at the University of the West of England successfully printed filigree chocolate designs", see Southerland D., Walters P., and Huson D. *Edible 3D printing.* (2011, January). In *NIP & Digital Fabrication Conference* (Vol. 2011, No. 2, pp. 819-822). Society for Imaging Science and Technology.

"This system was designed to allow customers to custom decorate their own cakes", see Sangen K. van der. *First 3D food printer in Dutch supermarket Albert Heijn.* (2015, February 9). 3DFoodPrinting.com. Retrieved March 11, 2016, from http://3dfoodprintingconference.com/food/first-3d-food-printer-dutch-supermarket-albert-heijn/

The information following *"A German company known as Biozoon has developed so-called "Smoothfood""* is from Pearse D. *Transforming mealtimes with 3D-printed food.* (2014, April 7). Horizon: The EU Research

& Innovation Magazine. Retrieved March 12, 2016, from http://horizon-magazine.eu/article/transforming-mealtimes-3d-printed-food_en.html

"3D printed pizza using multiple nozzles to print multiple ingredients in one composition", see Keating L. *You can now print edible pizza with a 3D 'Foodini' printer.* (2014, November 6). Tech Times. Retrieved April 11, 2016, from www.techtimes.com/articles/19630/20141106/startup-serves-up-edible-pizza-burgers-3d-foodini-printer.htm, accessed July 30, 2015.

"startups and companies have already discovered new ways to massively speed up the plastic 3D printing process", see Bourzac K. *Speeding Up 3-D Printing.* (2015, June 23). MIT Technology Review. Retrieved March 12, 2016, from https://www.technologyreview.com/s/538326/speeding-up-3-d-printing/

The information following the text *"When Jack Dorsey was sitting in a TechShop facility in California in 2008"* is based on the description in Hatch, M. (2013). *The maker movement manifesto: rules for innovation in the new world of crafters, hackers, and tinkerers.* McGraw Hill Professional.

The quote *"WiFi enabled, multi-color, energy efficient LED light bulb that can be controlled with an iPhone or Android."* is from *LIFX: The Light Bulb Reinvented.* (2013, December 11). Kickstarter. Retrieved March 12, 2016, from https://www.kickstarter.com/projects/lime-mouse/lifx-the-light-bulb-reinvented

"frustration at not being able to turn a light on and off because the switch was inaccessible.", see Williams P. *Product Innovation in a hyper connected world. The Australian Maker Movement.* (2014, May 29). Deloitte. Retrieved March 12, 2016, from http://www2.deloitte.com/au/en/pages/technology/articles/product-innovation-hyper-connected.html

"the fair has grown exponentially and in 2012, it attracted a crowd of 120,000 attendees", see *Maker Faire Bay Area 2012: Highlights and Headlines.* (2012, May 20). On3DPrinting. Retrieved March 12, 2016, from http://on3dprinting.com/2012/05/20/maker-faire-bay-area-2012-highlights-headlines/

"It featured over 100 Makers from more than 25 states and included more than 30 exhibits", see *A Nation of Makers.* WhiteHouse. Retrieved March 12, 2016, from https://www.whitehouse.gov/nation-of-makers

Chapter 4—Continuously Amplify Your Innovation A Million Fold

Rob McEwen's background comes from Macklem K. *Robert McEwen (Profile).* (2013, October 12). Historica Cananda. Retrieved March 15, 2016, from http://www.thecanadianencyclopedia.ca/en/article/robert-mcewen-profile/

"The nearby Campbell Lake mine had yielded an impressive 283 tons of gold (10 million ounces)", see Tischler L. *He Struck Gold on the Net (Really).* (2002, May 31). Fast Company. Retrieved March 15, 2016, from

http://www.fastcompany.com/44917/he-struck-gold-net-really

"with a market capitalization of around 100 million in 1999", see Goldcorp's 2000 Annual Shareholder Report, Goldcorp Inc. Management's Discussion & Analysis 2000 Consolidated Financial Statements.

The text following *"seminar for young presidents at the Massachusetts Institute of Technology"*, is based on Clayton G. Goldcorp's Virtual Prospecting Innovation. Institute for Global Entrepreneurship and Electronic Commerce. Retrieved March 15, 2016, from http://www.camese.org/uploads/goldcorp-fin.PDF

"only 100 million people in the United States (the country with the highest number of internet users at the time) were even connected to the Internet", see Internet live stats. *United States Internet Users*. Retrieved March 15, 2016, from http://www.internetlivestats.com/internet-users/united-states/

"as many countries as there are in the so-called "developed world", see Developed country. (2016, March 7). In *Wikipedia, The Free Encyclopedia*. Retrieved March 14, 2016, from https://en.wikipedia.org/w/index.php?title=Developed_country&oldid=708700785

"increased the worth of its company tenfold from 1999 to 2004", see Goldcorp share price which was $1.50 on November 1997, $2.25 on July 9th 1999, and then $17.9 on November 28, 2003.

"The campaign won Goldcorp much fame and recognition from the business community", see Clayton G. Goldcorp's Virtual Prospecting Innovation. Institute for Global Entrepreneurship and Electronic Commerce. Retrieved March 15, 2016, from http://www.camese.org/uploads/goldcorp-fin.PDF

The quote *"The act of a company or institution taking a function once performed by employees and outsourcing it to an undefined (and generally large) network of people in the form of an open call"* is from Howe J. *The Rise of Crowdsourcing*. (2006) Wired. Retrieved March 15, 2016, from http://www.wired.com/wired/archive/14.06/crowds.html

"a million readers in a major newspaper such as the New York Times or LA Times" Sicha C. *A Graphic History of Newspaper Circulation Over the Last Two Decades*. (2009, October 26). The Awl. Retrieved March 15, 2016, from http://www.theawl.com/2009/10/a-graphic-history-of-newspaper-circulation-over-the-last-two-decades

"It attracted over 150,000 people from 104 countries and 67 companies", see IBM. *IBM Invests $100 Million in Collaborative Innovation Ideas*. (2006, November 14). Retrieved March 15, 2016, from http://www-03.ibm.com/press/us/en/pressrelease/20605.wss

"In fact, around one million tasks are completed by the crowd each day", see Blattberg E. *How enterprses use crowdsourcing (infographic)*. (2013, November 14). Ven-

tureBeat. Retrieved March 15, 2016, from http://venture-beat.com/2013/11/14/how-enterprises-use-crowdsourc-ing-infographic/

The quote *"innovating with partners by sharing risk and sharing reward."* is from 100 Open. *Open Innovation Defined.* Retrieved March 15, 2016, from http://www.100open.com/2011/03/open-innovation-de-fined/

"the people solving your problems are experts", for de-mographics and motivation of the people solving crowdsourcing problems see Paolacci G., Chandler J. and Ipeirotis P. G. *Running Experiments on Amazon Me-chanical Turk.* 2010. Judgement and Decision Making 5 (5):411-19.

The quotes *"Our industry has been going through a hard time ... We had been trying to raise venture capital. Any positive news could only be a big help for us."* and *"It would have taken us years to get the recognition in North America that this project gave us."* are from Idea-Connection. *Open Innovation: Goldcorp Challenge.* Re-trieved March 15, 2016, from http://www.ideaconnec-tion.com/open-innovation-success/Open-Innovation-Goldcorp-Challenge-00031.html

The quote *"PRIZE WINNERS WILL BE REQUIRED TO EXECUTE AN ASSIGNMENT OF RIGHTS IN ORDER TO CLAIM THE PRIZE."* is from GrabCad. *GE Jet Engine Bracket Challenge – Official Rules.* Retrieved March 15, 2016, from http://blog.grabcad.com/ge-terms-of-ser-vice/

"Companies have thus learnt that breaking a complex problem down into more manageable parts produces better results", see King A. and Lakhani K. R. *Using Open Innovation to Identify the Best Ideas.* (2013 Fall). MITSloan Management Review. Retrieved March 15, 2016, from http://sloanreview.mit.edu/article/using-open-innovation-to-identify-the-best-ideas/

"lowering the barriers to entry and many other aspects need to be fine-tuned before one can expect a successful campaign", for the many aspects concerning this see Nielsen, M. *Reinventing discovery: the new era of networked science.* (2012). Princeton University Press.

"it was found that although around 30% of companies use inbound open innovation, only about 5% use outbound open innovation", see Schroll A. *First results from the open innovation study.* Alexander Schroll Website & Blog. Retrieved March 15, 2016, from http://www.schroll.eu/blog-open-innovation/125-first-results-from-the-open-innovation-study

The quote *"sells, licenses, or donates them to the external market."* is from Skarzynski, P., & Gibson, R. *Innovation to the core.* (2008). Harvard Business School Press, Boston.

The quotes *"in targeting diseases of the developing world—where there is not the same potential commercial return as in developed countries"* and *"by sharing expertise, resources, intellectual property and know-how with external researchers and the scientific community"* are from GSK. *Open Innovation.* Retrieved March

15, 2016, from http://uk.gsk.com/en-gb/research/shar-ing-our-research/open-innovation/

"taking out a staggering 7,481 patents", see Kharkovski R. *IBM leads in US patents for 22 years in a row.* (2015, January 19). IBM Advantage Blog. Retrieved March 15, 2016, from http://ibmadvantage.com/2015/01/19/ibm-leads-in-us-patents-for-22-years-in-a-row/

The story about James Gamble and William Procter is based on the history of Procter and Gamble provided by P&G entitled *Our History – How it began* at http://www.pg.com/en_US/downloads/me-dia/Fact_Sheets_CompanyHistory.pdf

"Ivory (1880s), Crisco (1911), Tide (1946), Prell (1947), Pampers (1961), Ariel (1967), Pert Plus (1986), and more", see Procter and Gamble's sheet *Innovation is P&G's Life Blood* available at https://www.pg.com/en_US/downloads/me-dia/Fact_Sheets_Innovation.pdf.

"P&G expanded globally, now operating in over 90 per-cent of the world, producing over 300 brands", see Procter and Gamble. *Connect + develop.* P&G. Retrieved March 15, 2016, from http://www.pgconnectde-velop.com/

"Procter & Gamble suffered from periods of lower than expected sales growth", see Skarzynski, P., & Gibson, R. *Innovation to the core.* (2008). Harvard Business School Press, Boston

The quote *"innovation is everyone's job"* and the follow-ing is from Lafley, A. G., & Charan, R. (2010). *The game changer: How every leader can drive everyday innovation*. Profile Books.

"more than 2,000 agreements around the world", see Procter and Gamble. *Connect + develop*. P&G. Retrieved March 15, 2016, from http://www.pgconnectde-velop.com/

"P&G slash its own R&D financial commitment by around 20%", see Skarzynski, P., & Gibson, R. *Innovation to the core*. (2008). Harvard Business School Press, Boston

"almost half of the ideas in its pipeline were attributed to its Connect and Develop program", see Huston L. and Sakkab N. *Connect and Develop: Inside Procter and Gamble's New Model for Innovation*. (2006, March). Harvard Business Review.

"The Aquafresh Isoactive", see Maven. *An Expert Perspective on Open Innovation*. (2012, March 29). Retrieved March 15, 2016, from http://www.maven.co/2012/03/29/an-expert-perspective-on-open-innovation

The quote *"seeks out and invests in early-stage ideas that could potentially revolutionize the energy industry"* is from Conser R. *Shell GameChanger – A Safe Place to Get Crazy Ideas Started*. (2013, January 7). Management Exchange. Retrieved March 15, 2016, from http://www.managementexchange.com/story/shell-game-changer

"GameChanger has partnered with over 1700 innovators", see Shell. *Shell GameChanger*. Retrieved March 15,

2016, from http://www.shell.com/global/future-energy/innovation/innovate-with-shell/shell-game-changer.html

"*more than $300 million in over 3000 ideas, turning about 250 of them into reality*", see Shell. *Innovation at Shell*. Retrieved March 15, 2016, from http://southafrica.shell.com/future-energy/innovation.html

"*As the latest research shows, this is a critical factor in more powerful innovation*", see Jorge. *Serendipitous exchanges fuel innovation*. (2015, August 21). Game-Changer. Retrieved March 15, 2016, from http://www.game-changer.net/2015/08/21/serendipitous-exchanges-fuel-innovation/

The quote "*allowing visitors to take existing video clips and music, insert their own words and create a customized 30-second commercial for the 2007 Chevy Tahoe*" is from Bosman, J. *Chevy Tries a Write-Your-Own-Ad Approach, and the Potshots Fly*, (2006, 4 April). New York Times. Retrieved March 15, 2016, from http://www.nytimes.com/2006/04/04/business/media/04adco.html?ex=1301803200&en=280e20c8ba110565&ei=5088&partner=rssnyt&emc=rss

The quote "*skewer[ed] everything from SUVs to Bush's environmental policy to ... the American automotive industry*" is from Howe, J. '*Neo Neologisms*', *Crowdsourcing: Tracking the Rise of the Amateur*. (2006, 16 June). Retrieved March 15, 2016, from http://crowdsourcing.typepad.com/cs/2006/06/neo_neologisms.html

"Flickr so they could sell photos uploaded by users", see
Dent S. *How to stop Yahoo from cashing in on your
Flickr images.* (2014, January 12). Engadget. Retrieved
March 15, 2016, from http://www.en-
gadget.com/2014/12/01/yahoo-creative-commons-flickr-
images/

The story about Threadless follows a number of sources
including Brabham D. C. *Crowdsourcing as a model for
problem solving an introduction and cases.* Conver-
gence: the international journal of research into new media technologies, 14(1), pp. 75—90, 2008, Sage Publications **as well as** Burkitt L. *Need To Build A Community?
Learn From Threadless.* (2010, July 1). Forbes. Retrieved
March 15, 2016, from
http://www.forbes.com/2010/01/06/threadless-t-shirt-
community-crowdsourcing-cmo-network-threadless.html
as well as Chafkin M. *The Customer is the Company.*
(2008, June 1). Inc. Retrieved March 15, 2016, from
http://www.inc.com/magazine/20080601/the-cus-
tomer-is-the-company_pagen_6.html

*"list of hundreds of companies trying to perfect the open
innovation business model and to provide people with a
platform on which to innovate"*, see Crowdsourcing.org.
2011 "Crowdsourcing Industry Landscape" [Infographic].
Retrieved March 15, 2016, from http://www.crowdsourc-
ing.org/document/2011-crowdsourcing-industry-land-
scape-infographic-/4111

"open innovation can quickly become open humiliation",
see Ehsani E. *Open Innovation and the BP oil spill: What
went wrong?* InnovationManagement.se. Retrieved

March 15, 2016, from http://www.innovationmanagement.se/imtool-articles/open-innovation-and-the-bp-oil-spill-what-went-wrong/

Chapter 5—Choosing A Winner The 21st Century Way

"The story above is an example of a famous experiment called the "dollar auction"", see Staw B. M. *The escalation of commitment: An update and appraisal.* (1997) In Shapira, Zur. Organizational Decision Making. New York, NY: Cambridge University Press. pp. 191–215.

"has been played all over the world with the above results coming up repeatedly", see Drummond H. *Are Good Leaders Decisive?* (1991). Management Decision, Vol. 29 Iss: 7, doi: http://dx.doi.org/10.1108/EUM0000000000075

"Try $2,000 and $1,950", see Murnighan J. K. *A very extreme case of the dollar auction.* (2002) Journal of Management Education, 26, pp. 56–69

"a landmark paper was published by Daniel Kahneman and Amos Tversky", see Kahneman D. and Tversky A. *Prospect theory: An analysis of decision under risk.* (1979) Econometrica: Journal of the Econometric Society, 263-291.

The quote *"for having integrated insights from psychological research into economic science, especially concerning human judgment and decision-making under uncertainty"* is from *The Sveriges Riksbank Prize in Economic Sciences in Memory of Alfred Nobel 2002.* Nobelprize.org. Retrieved March 31, 2016, from http://www.nobelprize.org/nobel_prizes/economic-sci-ences/laureates/2002/

The text following *"One of the most famous teams is in the United Kingdom, run by David Halpern"* comes from Rutter T. *The rise of nudge – the unit helping politicians to fathom human behaviour.* (2015, July 23). The Guardian. Retrieved March 31, 2016, from http://www.theguardian.com/public-leaders-net-work/2015/jul/23/rise-nudge-unit-politicians-human-behaviour

"they were some of the brightest university students at the world's top universities", see Ariely D. *Predictably Irrational.* (2008). Harper Collins. ISBN 978-0-06-135323-9.

"In fact, most seems an understatement because the facts point to 92% of them stumbling", see Marmer M., Herrmann B., Dogrultan E. and Berman R. Startup Genome Report Extra on Premature Scaling. (2012, March) Startup Genome. Retrieved March 31, 2016, from https://s3.amazonaws.com/startupcompass-pub-lic/StartupGenomeReport2_Why_Startups_Fail_v2.pdf

"It has been shown", see Booth B. *Correlation's Fresh Look At Venture Capital Returns.* (2013, November 18). Forbes. Retrieved March 31, 2016, from http://www.forbes.com/sites/bruce-booth/2013/11/18/correlations-fresh-look-at-venture-capital-returns/

"*when the fund plummets and returns a negative yield*", see Gigerenzer G. *Risk Savvy: How To Make Good Decisions*. (2014). Penguin.

"*to investigate the quality of expert judgments of high-stakes, real-world events*", see Tetlock P.E. *Expert political judgment: How good is it? How can we know?*. (2006). Princeton University Press.

"*illusory superiority*", see Hoorens V. *Self-enhancement and Superiority Biases in Social Comparison*. (1993). European Review of Social Psychology. Psychology Press. 4 (1): 113–139

"*90% of respondents claiming they were above-average drivers*", see Svenson O. *Are We All Less Risky and More Skillful Than Our Fellow Drivers?* (February 1981). Acta Psychologica 47 (2): 143–148. doi:10.1016/0001-6918(81)90005-6

"*which are extremely necessary in such a highly uncertain industry*", see Tetlock P.E. *Expert political judgment: How good is it? How can we know?*. (2006). Princeton University Press.

"*Over the long term, it will lead to more accurate decisions, which means less risk and more money left over for further investments*", see Gigerenzer G. *Risk Savvy: How To Make Good Decisions*. (2014). Penguin.

The text following "*During the 1980s*" comes from Chapman M. R. *In Search of Stupidity: Over Twenty Years of High Tech Marketing Disasters*. (2003). Apress.

"the most successful species are those which provide each other with mutual aid and support", see the work done by Peter Kropotkin. One can start with his Wikipedia page and go from there: Peter Kropotkin. (2016, March 20). In *Wikipedia, The Free Encyclopedia*. Retrieved March 31, 2016, from https://en.wikipedia.org/w/index.php?title=Peter_Kropotkin&oldid=711103464

"to under 5% five years later and by then they had become a takeover target", see Yanofsky D. *The smartphone business Microsoft bought in three charts*. (2013, September 3). Quartz. Retrieved March 31, 2016, from http://qz.com/120917/the-smartphone-business-microsoft-bought-in-three-charts/

"by giving people a break from the peer and social pressures of the group dynamics", see Epstein R. *Creativity for Crisis*. Drrobertepstein.com. Retrieved March 31, 2016, from http://drrobertepstein.com/downloads/CREATIVITY_FOR_CRISES-e-booklet-c_2009-Dr._Robert_Epstein.pdf?lbisphpreq=1

"over 100 other biases documented to date", see *The big list of behavioral biases*. The Psy-Fi Blog. Retrieved March 31, 2016, from http://www.psyfitec.com/p/the-big-list-of-behavioral-biases.html

"we have are our intuitions and heuristic thinking (hence biases!)", and for the quote *"Calculated intelligence may do the job for known risks, but in the face of uncertainty, intuition is indispensable"*, see Gigerenzer G. *Risk Savvy: How To Make Good Decisions*. (2014). Penguin.

The quote *"the issues that matter"* is from Kahneman D. and Klein G. *Conditions for Intuitive Expertise: A Failure to Disagree.* (September 2009). American Psychologist 64, no. 6, pp. 515– 26.

Chapter 6—Free Knowledge For Free Innovation

"the Journal of Comparative Neurology have subscription fees of around $30,000 for both the print and online publication", see Dobbs D. *The head of the Harvard library system is pissed.* (2010, December 17). Wired. Retrieved April 25, 2016, from http://www.wired.com/2010/12/the-head-of-the-harvard-library-system-is-pissed/

"No Rights Reserved", see Prabhala A. *Would Gandhi have been a Wikipedian?* (2012, January 17). The Indian Express. Retrieved March 13, 2016, from http://archive.indianexpress.com/news/would-gandhi-have-been-a-wikipedian/900506/0

Much of the information following *"The business of scholarly publishing is highly lucrative"* comes from Poynder R. *The Basement Interviews: Peter Suber.* (2007, October 19). Richard Poynder Blogspot. Retrieved March 13, 2016, from http://poynder.blogspot.com.au/2007/10/basement-interviews-peter-suber.html

"valued at ten billion dollars", see Esposito J. *A snapshot of the Scientific and Technical Publishing Market.* (2013, November 4). The Scholarly Kitchen. Retrieved March 13,

2016, from http://scholarlykitchen.ssp-net.org/2013/11/04/a-snapshot-of-the-scientific-and-technical-publishing-market/

"the first woman slated by the American Democratic Party for state-wide office", see State of Illinois, 90th General Assembly Legislation. Retrieved March 13, 2016, from http://www.ilga.gov/legisla-tion/legisnet90/srgroups/sr/900SJ0072LV.html

"Nine years later, that figure grew to around 200,000", see Laakso M., Welling P., Bukvova H., Nyman L., Björk B. C., and Hedlund, T. (2011). Hermes-Lima, Marcelo, ed. *The Development of Open Access Journal Publishing from 1993 to 2009. PLoS ONE* **6** (6): e20961. doi:10.1371/journal.pone.0020961. PMC 3113847. PMID 21695139

"arXiv already archiving its one millionth article in 2014", Noorden R. V. *The arXiv preprint server hits 1 million articles.* (2014, December 30). Nature. Retrieved April 11, 2016, from http://www.nature.com/news/the-arxiv-preprint-server-hits-1-million-articles-1.16643

The quote *"free availability on the public internet..."* comes from *Read the Budapest Open Access Initiative.* Retrieved March 13, 2016, from http://www.buda-pestopenaccessinitiative.org/read

"By the end of 2015 this document would have attracted more than 5,900 individuals and over 800 organiza-tions", see *View Signatures.* Retrieved March 13, 2016, from http://www.budapestopenaccessinitia-tive.org/list_signatures

The quote *"offer distributed full-text access to all DOE-affiliated accepted manuscripts or articles after an administrative interval of 12 months"* is from Department of Energy. PAGES. Retrieved March 13, 2016, from http://www.osti.gov/pages/

The quote and information after *""deposited in a public access compliant repository designated by NSF." This program began in January 2016."* is from National Science Foundation, Public Access: Frequently Asked Questions. Retrieved March 24, 2016, from http://www.nsf.gov/pubs/2016/nsf16009/nsf16009.jsp#q1

"They are required to make peer reviewed journal articles available in the author's institutional repository or any other easily accessible location" and *"having adopted the practice in 2006"*, see Open access. (2016, March 12). In *Wikipedia, The Free Encyclopedia*. Retrieved March 13, 2016, from https://en.wikipedia.org/w/index.php?title=Open_access&oldid=709617846

"Tri-Agency Open Access Policy", see Government of Canada. *Tri-Agency Open Access Policy on Publications*. (2015, February 27). Retrieved March 13, 2016, from http://www.science.gc.ca/default.asp?lang=En&n=F6765465-1

"open access publications increase the number of citations a work receives", see Hajjem C., Harnad S. and Gingras Y. *Ten-Year Cross-Disciplinary Comparison of the Growth of Open Access and How It Increases Research Citation Impact.* (2005). *IEEE Data Engineering*

Bulletin 28 (4): 39–47. The article analyzes 1,307,038 articles published across 12 years (1992–2003) in 10 disciplines; OA articles have consistently more citations (25%–250% varying with discipline and year).

 "40,000 Americans were dying from a particular form of cancer", see *What are the key statistics about pancreatic cancer?* (2015, January 9). American Cancer Society. Retrieved March 13, 2016, from http://www.cancer.org/cancer/pancreaticcancer/detailedguide/pancreatic-cancer-key-statistics

"allegedly aroused by the death of a close family friend", see *Intel Science Winner Develops Cancer Tech.* (2012, May 23). WSJ Video. Retrieved March 13, 2016, from http://live.wsj.com/video/intel-science-winner-develops-cancer-tech/E342B43B-F184-492D-A441-38B28C18D3C1.html#!E342B43B-F184-492D-A441-38B28C18D3C1

"100 times faster, tens of thousands dollars cheaper, and is hundreds of times more sensitive than other leading tests." The young scientist claims are impressive and his goals laudable but a one-man show is always difficult. Although his ambitions are exemplary, there have been questions over his methodology and his cost, speed and sensitivity claims, see for example Sharon E., Zhang J., Hollevoet K., Steinberg S. M., Pastan I., Onda M., Gaedcke J., Ghadimi B. M., Ried T., Hassan R. *Serum mesothelin and megakaryocyte potentiating factor in pancreatic and biliary cancers.* Clin Chem Lab Med. 2012 April;50(4):721-5. doi: 10.1515/CCLM.2011.816. [1],

PMID 22149739. In addition, although he made his discovery utilizing open forms of information he has been criticized for not openly publishing his discovery and, for some an even worse crime, filing a patent for it.

"began his own independent research into immunology and gene therapy", see *Matthew Scholz*. CrunchBase. Retrieved March 13, 2016, from https://www.crunchbase.com/person/matthew-scholz

The text following *"To make significant progress in cancer research"* is based on Timmerman L. A Computer Guy's Dream, Immusoft, Turns Cells Into Drug Factories. (2012, July 5). Xconomy. Retrieved March 13, 2016, from http://www.xconomy.com/seattle/2012/07/05/a-computer-guys-dream-immusoft-turns-cells-into-drug-factories/

"it was made possible by open access journals and Wikipedia", see Regalado A. *A Contrarian in Biotech*. (2015, March 16). MIT Technology Review. Retrieved April 11, 2016, from http://www.technologyreview.com/news/535771/a-contrarian-in-biotech/

"the second leading cause of death in America", see *Leading Causes of Death*. (2016, February 25). Centers for Disease Control and Prevention. Retrieved March 13, 2016, from http://www.cdc.gov/nchs/fastats/leading-causes-of-death.htm

"Immusoft has raised $2.7 million", see *Immusoft*. Crunchbase. Retrieved March 13, 2016, from https://www.crunchbase.com/organization/immusoft

"*and had successfully completed tests on mice*", see Geddes. L. *Mini drug factory churns out drugs from inside bone*. (2013, September 21). New Scientist. Retrieved April 11, 2016, from https://www.newscientist.com/article/mg21929352.600-mini-drug-factory-churns-out-drugs-from-inside-bone/#.VX82yFVVhBc

The quote "*The odds are against him probably getting this far*" is from Timmerman L. A *Computer Guy's Dream, Immusoft, Turns Cells Into Drug Factories*. (2012, July 5). Xconomy. Retrieved March 13, 2016, from http://www.xconomy.com/seattle/2012/07/05/a-computer-guys-dream-immusoft-turns-cells-into-drug-factories/

"*edict by Emperor Theodosius I in 391 made paganism illegal*", see Norwich J. J. *Byzantium: The Early Centuries. (1988)*. Viking.

"*many budgets have remained stagnant or even been cut*", there exist many articles on this. See, for example, *Library Budget Cuts*. The Huffington Post. Retrieved March 13, 2016, from http://www.huffingtonpost.com/news/library-budget-cuts/

"*increase by almost ten percent annually*", see Serials Price Projections for 2015. EBSCO. Retrieved March 13, 2016, from https://www.ebscohost.com/promoMaterials/Serials_Price_Projections_for_2015.pdf

"*Luckily, a group of audacious researchers recently made details of these contracts public*", see Bergstrom T. C., Courant P. N., McAfee R. P. and Williams M. A. *Eval-*

uating big deal journal bundles. Proceedings of the National Academy of Sciences, Vol 111, No. 26, pp. 9425–9430, 2014, National Acad Sciences.

"Reed Elsevier Group alone spent over $23 million to lobby the American government", see Reed Elsevier Group. (2016, January 22). OpenSecrets.org. Retrieved March 13, 2016, from http://www.opensecrets.org/lobby/clientsum.php?id=D000067394&year=2015

"The currently available 10,000 journals", see Morrison H. *Dramatic Growth of Open Access September 30, 2014: some useful numbers for open access week*. (2014, October 1). Heather Morrison Blog. Retrieved March 13, 2016, from http://poeticeconomics.blogspot.com.au/2014/10/dramatic-growth-of-open-access.html

The quote *"the century of physics"* comes from Venter C. and Cohen D. *The Century of Biology*. (2004, November 1). New Perspectives Quarterly, 21: 73–77. doi: 10.1111/j.1540-5842.2004.00701.x

The facts following *"It manages the largest particle physics laboratory in the world"* comes from http://public-archive.web.cern.ch/public-archive/en/About/Nobels-en.html **and** http://lhc-machine-outreach.web.cern.ch/lhc-machine-outreach/lhc-interesting-facts.htm

"It alone receives $1.3 billion per year", see Knapp A. *How Much Does It Cost To Find A Higgs Boson?* (2012, July 5). Forbes. Retrieved March 13, 2016, from

http://www.forbes.com/sites/alexknapp/2012/07/05/ho
w-much-does-it-cost-to-find-a-higgs-boson/

*"only behind projects such as the International Space
Station and the Apollo Program"*, see Wåhlberg O. *I
Vetenskapens Värld på måndag – De fem dyraste
vetenskapsprojekten någonsin.* (2010, April 15). Svt
Nyheter. Retrieved March 13, 2016, from
http://www.svt.se/nyheter/vetenskap/i-vetenskapens-
varld-pa-mandag-de-fem-dyraste-vetenskapsprojekten-
nagonsin

*"with the deepest part almost 200m below the earth's
surface" and "the information is sent around the world"*,
see Large Hadron Collider. (2016, March 13). In *Wikipedia, The Free Encyclopedia.* Retrieved March 13, 2016,
from https://en.wikipedia.org/w/index.php?ti-
tle=Large_Hadron_Collider&oldid=709785673

The quote *"When you are working with 3,000 collabora-
tors from across the world, open access certainly does
make life an awful lot easier. In fact, I don't think we
could do the work we do without it"* comes from Professor Tony Doyle, CERN Atlas. *Open Access Success Stories.* Retrieved March 13, 2016, from http://www.oasto-
ries.org/2011/09/ukworldwide-researcher-professor-
tony-doyle-cern-atlas/

"the Open Data Portal", see O'Luanaugh C. *CERN makes
public first data of LHC experiments.* (2014, November
20). CERN. Retrieved March 13, 2016, from
http://home.web.cern.ch/about/updates/2014/11/cern-
makes-public-first-data-lhc-experiments

"60 articles circulating amongst researchers and being discussed for their implications", see Professor Tony Doyle, CERN Atlas. *Open Access Success Stories*. Retrieved March 13, 2016, from http://www.oastories.org/2011/09/ukworldwide-researcher-professor-tony-doyle-cern-atlas/

"he won the prestigious ArsDigita Prize for creating The Info Network", see both Aaron Swartz. Retrieved March 13, 2016, from http://knowyourmeme.com/memes/people/aaron-swartz **and** Schofield J. *Aaron Swartz obituary*. (January 13, 2013). The Guardian (London). Retrieved March 13, 2016, from http://www.theguardian.com/technology/2013/jan/13/aaron-swartz

"The movement garnered the support of over 115,000 websites", see Wortham J. *Public Outcry Over Antipiracy Bills Began as Grass-Roots Grumbling*. (2012, January 19). New York Times. Retrieved April 11, 2016, from http://www.nytimes.com/2012/01/20/technology/public-outcry-over-antipiracy-bills-began-as-grass-roots-grumbling.html?pagewanted=1&ref=technology&_r=0

"He did this by way of a Massachusetts Institute of Technology JSTOR journal archive account" and *"he was charged with wire and computer fraud"*, see "Indictment, USA v. Swartz, 1:11-cr-10260, No. 2 (D.Mass. July 14, 2011)". MIT. July 14, 2011. Superseded by "Superseding Indictment, USA v. Swartz, 1:11-cr-10260, No. 53 (D.Mass. September 12, 2012)". Docketalarm.com. September 12, 2012.

"a passionate plea urging people to "fight back" against the "private theft of public culture"", see Swartz A. *Guerilla Open Access Manifesto*. (2008 July). Retrieved March 13, 2016, from https://archive.org/stream/GuerillaOpenAccessManifesto/Goamjuly2008_djvu.txt

The quote *"He used his prodigious skills as a programmer and technologist not to enrich himself but to make the Internet and the world a fairer, better place."* is from Nelson V. J. *Aaron Swartz dies at 26; Internet folk hero founded Reddit*. (January 12, 2013). Los Angeles Times. Retrieved March 13, 2016, from http://www.latimes.com/news/obituaries/la-me-0113-aaron-swartz-20130113,0,5232490.story

"The paper was accepted by a shocking 157 journals", see Bohannon J. *Who's Afraid of Peer Review*. (2013, October 4). Science. Retrieved March 13, 2016, from http://www.sciencemag.org/content/342/6154/60.summary

The quote *"[a]ny reviewer with more than a high-school knowledge of chemistry and the ability to understand a basic data plot should have spotted the paper's shortcomings immediately"* also comes from Bohannon J. *Who's Afraid of Peer Review*. (2013, October 4). Science. Retrieved March 13, 2016, from http://www.sciencemag.org/content/342/6154/60.summary

The quote *"[Gold] Open Access will likely make it easier to consume research outputs, but harder to produce them."* comes from Kostoulas A. *Open Access: Some*

Facts and Some Thoughts. (2013 October 23). Achille-asKostoulas.com. Retrieved March 13, 2016, from http://achilleaskostoulas.com/2013/10/23/open-access-some-facts-and-some-thoughts/

The quote *"not identify any losses of subscriptions for this reason and that they do not view arXiv as a threat to their business ... rather the opposite"* comes from Swan A. *Open access self-archiving: An Introduction.* (2005, June 19). University of Southhampton. Retreived March 13, 2016, from http://eprints.soton.ac.uk/261006/

Chapter 7—The Innovation Services Revolution

"$2 billion of venture capital pouring into cloud computing in the United States each year since 2011", see slide 72 in *The Future of Cloud Computing*. North Bridge. Retrieved March 10, 2016, from http://www.north-bridge.com/2013-cloud-computing-survey

"In 2015, that had risen to more than $170 billion", see slide 6 in Deeter B. *State of the Cloud Report*. Bessemer Venture Partners. Retrieved March 10, 2016, from http://clearslide.com/view/mail?iID=UGVYTY42F2KZH XLV54B7

"Seen in this light, cloud-based business models are not new.", the following text draws some of its facts from Software as a service. (2016, March 5). In *Wikipedia, The Free Encyclopedia*. Retrieved March 10, 2016, from https://en.wikipedia.org/w/index.php?title=Software_as_a_service&oldid=708416639

John McCarthy's quote *"may someday be organized as a public utility"* comes from Schaefer D. *Cloud-Based Design and Manufacturing (CBDM): A Service-Oriented Product Development Paradigm for the 21st Century*, Springer; 2014 edition

"the Software as a Service sector will be worth more than $US100 billion", see Columbus L. *Roundup of Cloud Computing Forecasts and Market Estimates*. (2015, January 24). Forbes. Retrieved March 10, 2016, from http://www.forbes.com/sites/louiscolumbus/2015/01/24/roundup-of-cloud-computing-forecasts-and-market-estimates-2015/#616f5c4740ce

"the size of the global toy market", see Statista. *Statistics and facts on the Toy Industry*. Statista. Retrieved March 10, 2016, from http://www.statista.com/topics/1108/toy-industry/

"Between 2008 and 2015, the entire "Anything as a Service"", see Columbus L. *Roundup of Cloud Computing Forecasts and Market Estimates*. (2015, January 24). Forbes. Retrieved March 10, 2016, from http://www.forbes.com/sites/louiscolumbus/2015/01/24/roundup-of-cloud-computing-forecasts-and-market-estimates-2015/#616f5c4740ce

"The primary reason? Increased agility", see Deans D. H. *How cloud adoption trends are driven by strategic imperatives*. (2014, December 22). CloudTech. Retrieved March 10, 2016, from http://www.cloudcomputing-news.net/news/2014/dec/22/cloud-adoption-trend-is-driven-by-strategic-imperatives/

"Everything from accounting to client relationship, management to marketing and even healthcare solutions", see Columbus L. *America's Fastest Growing Enterprise Software Companies of 2013*. (2013, November 30). Forbes. Retrieved March 10, 2016, from http://www.forbes.com/sites/louiscolumbus/2013/11/30/americas-fastest-growing-enterprise-software-companies-of-2013/#2f1d64516694

"it was estimated that Amazon's cloud revenue had eclipsed the combined cloud revenues of Google, IBM, Microsoft and Salesforce.com", see Darrow B. *The battle for cloud supremacy: Amazon's AWS vs. legacy IT juggernauts*. (2015, April 27). Fortune. Retrieved March 10, 2016, from http://fortune.com/2015/04/27/amazon-aws-cloud/

"already making up one sixth of the Internet booksellers' revenue, and rising", see Lee N. *Amazon Web Services is a $5 billion business*. (2015, April 23). Engadget. Retrieved March 10, 2016, from http://www.engadget.com/2015/04/23/amazon-q1-2015/

Bezos' quote *"We want to make money when people use our devices, not when they buy our devices"* comes from Niu E. *How Amazon's Hardware-as-a-Service Strategy Is a Major Disruptor*. (2012, September 12). The Motley Fool. Retrieved March 10, 2016, from http://www.fool.com/investing/general/2012/09/12/how-amazons-hardware-as-a-service-strategy-is-a-ma.aspx?source=isesitlnk0000001&mrr=1.00

"It only has industries like government and finance ahead of it", see House J. *Five Takeaways From New GDP-by-Industry Report.* (2014, April 25). The Wall Street Journal. Retrieved March 10, 2016, from http://blogs.wsj.com/economics/2014/04/25/five-takea-ways-from-new-gdp-by-industry-report/

The paragraphs following *"The idea for Shapeways took place in 2007"* are based on Wirth M. and Thiesse F. *SHAPEWAYS AND THE 3D PRINTING REVOLUTION.* (2014, June). Retrieved March 10, 2016, from http://cedifa.de/wp-content/up-loads/2014/06/ECIS2014_Shapeways_TeachingCase.pdf

"printing of its one-millionth item already in June 2012", see Lovecraft R. *Shapeways hits one million 3D printed creations.* (2012, June 20). *TG Daily.*

"to print toy characters from their shows", see Harris E. A. *Hasbro to Collaborate With 3-D Printing Company to Sell Artwork.* (2014, July 20). New York Times. Retrieved March 10, 2016, from http://www.ny-times.com/2014/07/21/business/hasbro-selling-my-lit-tle-pony-fan-art.html?_r=0

The quote *"users with the ability to utilize the manufac-turing capabilities of configurable, virtualized produc-tion networks, based on cloud-enabled, federated facto-ries, supported by a set of software-as-a-service applica-tions"* comes from the Manucloud Project. Retrieved March 10, 2016, from http://www.manucloud-pro-ject.eu/index.php?id=233.

The quote *"down to the shop floor level"* comes from Schaefer D. *Cloud-Based Design and Manufacturing (CBDM): A Service-Oriented Product Development Paradigm for the 21st Century*, Springer; 2014 edition

"even major American health insurance companies such as United Healthcare will cover these virtual visits", see Lapowsky I. *Video is about to become the way we all visit the doctor.* (2015, April 30). Wired. Retrieved March 10, 2016, from http://www.wired.com/2015/04/united-healthcare-telemedicine/

"Like one in five Americans, you probably own a so-called health wearable", see Comstock J. *PwC: 1 in 5 Americans owns a wearable, 1 in 10 wears them daily.* (2014, October 21). MobiHealthNews. Retrieved March 10, 2016, from http://mobihealthnews.com/37543/pwc-1-in-5-americans-owns-a-wearable-1-in-10-wears-them-daily/

"pioneering operations performed in Vancouver by the so-called Arthrobot", see Shadab M. *Robotic Surgery* (2013, March 27). Journal of Oral Biology and Craniofacial Research.

"With some machines costing millions of dollars", see Singer E. *The Slow Rise of the Robot Surgeon.* (2010, March 24). MIT Technology Review. Retrieved March 10, 2016, from https://www.technologyreview.com/s/418141/the-slow-rise-of-the-robot-surgeon/

"doctors have understandably questioned their cost-saving potential", see Kolata G. *Results Unproven, Robotic*

Surgery Wins Converts. (2010, February 13). The New York Times.

"in order to realize the dream of operating remotely around the world", see Lalwani M. *Telesurgery tests highlight the limits of the Internet*. (2015, May 5). Engadget. Retrieved March 10, 2016, from http://www.engadget.com/2015/05/05/telesurgery-tests-highlight-the-limits-of-the-internet/

Chapter 8—The Bigger Picture

"the number of Google searches for the term innovate doubled", see Google Trends. *Innovate*. Retrieved March 14, 2016, from https://www.google.com.au/trends/explore#q=innovate

"but also because it should lead into a cycle of more discoveries which succeed sooner." Of course this argument ignores the effort involved in managing many small investments over one large one but if the amount invested is big enough, then sometimes one large investment can cause more work and effort than many small ones combined. In any case, what this actually looks like in the real world may vary from case to case.

"John was devastated." John was not the person's real name although the details here are based on a real person.

The text following *"The whole ordeal begins with the discovery of a molecule"* is based on Barnes D. *How prescription drugs are developed*. (2006, December). Australian Prescriber. Retrieved March 14, 2016, from

http://m.australianprescriber.com/maga-
zine/29/6/159/61

The quote *"Consistently ranks as one of the most profitable industries in the United States"* is from Congressional
Budget Office. *Research and Development in the Pharmaceutical Industry.* (2006, October 1). Retrieved March
14, 2016, from https://www.cbo.gov/publication/18176

"commands up to $10 billion per year, in the U.S. alone",
see Drugs.com. *U.S. Pharmaceutical Sales – Q4 2013.*
(2014, February). Retrieved March 14, 2016, from
http://www.drugs.com/stats/top100/sales

*"At its peak, the new compound was earning its creators
between two and three billion dollars a year"* and
*"80,000 people suffering from heart attacks or strokes
as a result of taking Avandia"* and the quote *"The medicine increased heart attack risk by a whopping 43 percent"*, see Thomas C. *Avandia: a very short history of a
very bad drug.* (2013, January 21). Ethicalnag.org. Retrieved March 14, 2016, from http://ethical-
nag.org/2013/01/21/avandia-a-very-short-history-of-a-
very-bad-drug/

*"paid out over $3 billion in fines for "failing to report
drug safety information""*, see Sifferlin A. *Breaking
Down GlaxoSmithKline's Billion-Dollar Wrongdoing.*
(2012, July 5). Time. Retrieved March 14, 2016, from
http://healthland.time.com/2012/07/05/breaking-
down-glaxosmithklines-billion-dollar-wrongdoing/

The quotes *"escalating costs and the declining number of
launches"* and *"[W]ithout a substantial increase in R&D*

productivity, the pharmaceutical industry's survival (let alone its continued growth prospects), at least in its current form, is in great jeopardy." are from Paul S. M., Mytelka D. S., Dunwiddie C. T., Persinger C. C., Munos B. H., Lindborg S. R. and Schacht A. L. *How to improve R&D productivity: the pharmaceutical industry's grand challenge.* (2010 March). Nature Reviews. Retrieved March 14, 2016, from http://www.nature.com/nrd/journal/v9/n3/full/nrd3078.html

The text and quotes following *"The indication is then, that companies are getting fewer results for more research spending"* are from Congressional Budget Office. *Research and Development in the Pharmaceutical Industry.* (2006, October 1). Retrieved March 14, 2016, from https://www.cbo.gov/publication/18176

"disproportionate gain in the number of successful drugs is realized", see Figure 5 in Paul S. M., Mytelka D. S., Dunwiddie C. T., Persinger C. C., Munos B. H., Lindborg S. R. and Schacht A. L. *How to improve R&D productivity: the pharmaceutical industry's grand challenge.* (2010 March). Nature Reviews. Retrieved March 14, 2016, from http://www.nature.com/nrd/journal/v9/n3/full/nrd3078.html

The quote *"Technology is transforming innovation at its core, allowing companies to test new ideas at … prices that were unimaginable even a decade ago"* is from Brynjolfsson E. and Schrage M. *The New, Faster Face of Innovation.* (2009, August 17). The Wall Street Journal. Retrieved March 14, 2016, from http://www.wsj.com/ar-

ti-
cles/SB10001424052970204830304574130820184260340

Index

www.ingramcontent.com/pod-product-compliance
Lightning Source LLC
Chambersburg PA
CBHW060329200326
41519CB00011BA/1881